MEXICO
Total Eclipse Guide

Official Commemorative 2024 Keepsake Guidebook

2024 Total Eclipse Guide

Aaron Linsdau

Sastrugi Press

Jackson Hole

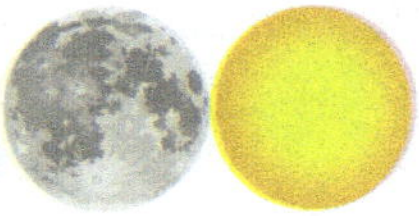

Sastrugi Press / Published by arrangement with the author
Mexico Total Eclipse Guide: Official Commemorative 2024 Keepsake Guidebook

The author has made every effort to accurately describe the locations contained in this work. Travel to some locations in this book is hazardous. The publisher has no control over and does not assume any responsibility for author or third-party websites or their content describing these locations, how to travel there, nor how to do it safely. Refer to local regulations and laws.

Any person exploring these locations is personally responsible for checking local conditions prior to departure. You are responsible for your own actions and decisions. The information contained in this work is based solely on the author's research at the time of publication and may not be accurate in the future. Neither the publisher nor the author assumes any liability for anyone climbing, exploring, visiting, or traveling to the locations described in this work. Climbing is dangerous by its nature. Any person engaging in mountain climbing is responsible for learning the proper techniques. The reader assumes all risks and accepts full responsibility for injuries, including death.

Sastrugi Press
PO Box 1297, Jackson, WY 83001, United States
www.sastrugipress.com
Quantity sales: Special discounts are available on quantity purchases by corporations, associations, and others. For details, contact the publisher at the address above.
Library of Congress Catalog-in-Publication Data
Library of Congress Control Number: 2019949789
Linsdau, Aaron
Mexico Total Eclipse Guide / Aaron Linsdau-1st United States edition
p. cm.
1. Nature 2. Astronomy 3. Travel 4. Photography
Summary: Learn everything you need to know about viewing, experiencing, and photographing the total eclipse in Mexico on April 8, 2024.

ISBN-13: 978-1-64922-332-6 (paperback)

508.4—dc23

All photography, maps and artwork by the author, except as noted.
00067

10 9 8 7 6 5 4 3

Contents

Introduction	4
All About Mexico	6
Overview of Mexico	6
Weather	11
Finding the Right Location	13
Road Closures and Traffic	14
Wilderness and Forest Park Safety	15
Eclipse Day Safety	17
All About Eclipses	19
Total vs Partial Eclipse	20
Early Myth & Astronomy	22
Contemporary American Solar Phenomena	24
Future American Eclipses	26
Viewing and Photographing the Eclipse	27
Planning Ahead	28
Understanding Sun Position	29
Eclipse Data for Selected Locations	31
Eclipse Photography	31
Eclipse Photography Gear	34
Camera Phones	36
Viewing Locations Around Mexico	43
Remember the Mexico Total Eclipse	65

Introduction

Thank you for purchasing this book. It has everything you need to know about the total eclipse in Mexico on April 8, 2024.

A total eclipse passing over central Mexico is a rare event. The last Mexican total eclipse was in 1991. It traveled the length of central Mexico. The last Mexican total eclipse prior to that was in 1970!

The next total eclipse over Mexico will not be until April 8, 2024. It will pass over Mexico, the United States, and on to Canada. After that, the next total eclipse in Mexico will be in 2052.

It's imperative to make travel plans early. You will be amazed at the number of people swarming to the total eclipse path. Some might say watching a partial versus a total eclipse is a similar experience. It's not.

This book is written for Mexico visitors and anyone else viewing the eclipse. You will find general planning, viewing, and photography information inside. Should you travel to the eclipse path in Mexico in April, be prepared for an epic trip. The estimates based on the 2017 eclipse suggest that millions will converge on central Mexico.

Some hotels in the communities and cities along the path of totality in Mexico have already been contacted by people to make reservations. Finding lodging along the eclipse path may be a major challenge.

Resources will be stretched far beyond the normal limits. Think of the biggest tourist event possible and even bigger. It may be likely that traffic along highways will come to a complete standstill during this event. Be prepared with backup supplies.

Many smaller Mexican towns are far from any major city. Mexico country roads can be slow. Please obey posted speed limits for the safety of everyone. Be cautious about believing a map application's estimate of travel time in Mexico.

People in communities along the path of the total eclipse may rent out properties for this event. With this major celestial spectacle in the spring of 2024, be assured that Mexico "hasn't seen anything yet."

Is this to say to avoid Mexico or other areas during the eclipse? Not at all! This guidebook provides ideas for interesting, alternative, and memorable locations to see the eclipse. It will be too late to rush to a better spot once the eclipse begins. Law enforcement will be out to help drivers reconsider speeding.

Please be patient and careful. There will be a large rush of people from all over the world, converging on Mexico to enjoy the total eclipse. Be mindful of other drivers on eclipse weekend, as they may not be familiar with Mexican roads.

You should feel compelled to play hooky on April 8. Ask for the day off. Take your kids out of school. They'll be adults before the next chance to see a total eclipse over Mexico. Create family memories that will last a lifetime. Sastrugi Press does not normally advocate skipping school or work. Make an exception because this is too big an event to miss.

Wherever you plan to be along the total eclipse path, leave early and remember your eclipse glasses. People from all around the planet will converge on central Mexico. Be good to your fellow humans and be safe. We all want to enjoy this spectacular show.

Visit www.sastrugipress.com/eclipse for the latest updates for this state eclipse book series.

Author Information

Polar explorer and motivational speaker Aaron Linsdau's first book, *Antarctic Tears*, is an emotional journey into the heart of Antarctica. He ate two sticks of butter every day to survive. Aaron coughed up blood early in the expedition and struggled with equipment failures. Despite the endless difficulties, he set a world record for surviving the longest solo expedition to the South Pole.

Aaron teaches how to build resilience to overcome adversity by managing attitude. He shares his techniques for overcoming adrenaline burnout and constant overload. He inspires audiences to face their challenges with a new perspective. As a motivational speaker, Aaron talks about courage, resilience, attitude, safety, and risk. He hopes that you will be inspired and have an enjoyable time watching the total eclipse in Mexico. Visit www.aaronlinsdau.com/subscribe to sign up for his mailing list to learn more.

All About Mexico

Overview of Mexico

Mexico is a place where unique cultures converge across equally diverse landscapes giving rise to endless festivities, astonishing artistry, and rich history. Natural attractions vary from diverse ecosystems packed with exotic species to long stretches of uninterrupted wilderness. Rolling hills and clear waters are perfect for athletic endeavors. Its cities provide rest in the form of ancient churches, parks, and delicious cuisine. Mexico is the perfect destination for families and solo travelers alike for the total eclipse.

One of Mexico's most striking features is the sugar-white dunes just outside of Cuatro Ciénegas, Coahuila, are the Dunas de Yeso. The snowy white desert made of gypsum is nearly 2,000 acres (800 hectares). Where the white dunes end, more deserts begin. This makes the wetlands buried between its vast stretches of sand all the more surprising. The wetlands consist of several shimmering blue pools of water. Each pool and the surrounding habitats teem with extensive wildlife earning the location the title, "Galapagos of Mexico." Here, visitors may enjoy the hot springs, swimming, camping, or snorkeling.

Once you have exhausted the many sites around Cuatro Ciénegas, continue north into the mountain ranges. A range of mountains known as the Sierra de la Campaña surrounds the desert. This range provides visitors with a beautiful hiking zone from which the marshes and desert are both visible. The wetlands environment of Poza Azul is not to be missed.

Coahuila's beauty doesn't end above ground. Hidden below ground in the city of Torreón is El Canal de la Perla (the Pearl Watercourse). The tunnel leads tourists on a historical underground tour of the city lit by lanterns. The original canal was over 7.5 miles (12 km) long. Lost for decades, it was rediscovered in 2003 by a construction crew. Then there is the hard-to-miss Cristo de las Noas. A 69-foot (21-m) tall statue of Jesus spreads its arms out above the city below. It is a breathtaking sight at the edge of Coahuila that is named after the endangered small cactus that grows in the area.

El Paseo del Viejo Oeste is a western film set turned theme park located 12.6 miles (20.3 km) north of Durango. Visitors can ride ponies, take photos in period costumes, and witness shows full of cowboys, dancers, and Apache Indians. Not only are children able to enjoy a fantasy of living as a desperado, but everyone also receives an educational glimpse into a bygone era that has influenced both history and media.

To continue the walk back into history, you can visit Durango's Túnel Minería Museum. It hosts artifacts of Durango's ancient civilizations in displays 33 feet (10 m) below the city. Visitors are equipped with reflective vests and mining hats for total immersion. At the end of the tour, children pan for rocks to take home as souvenirs.

Sinaloa's most famous city, Mazatlán, has the famous Malecón to explore. The 12-mile (19.3-km) waterfront provides a scenic look at Mazatlán city. There is room aplenty for pedestrians, bikers, drivers, and those who choose the area's iconic pulmonia taxi. On the way, you'll see Mazatlán's many monuments including the Family monument, Pacifico Brewery monument, Fisherman's Monument, and the monument to the pulmonia taxi itself. In the afternoon, you can settle in on the beach to watch divers perform stunts from Punta de Clavadistas (El Clavadista Cliff).

Off the coast of Mazatlán is the Isla de Piedra (Stone Island), an island known for bird sighting such as toucans and parrots. Jet skis, surfboards, and banana boats are available to explore the waters surrounding the island. If you'd rather go farther out to sea, the Santa Maria Bay in Altamura offers boat trips and fishing expeditions. Another noteworthy location near Mazatlán is Cerro Creston, the hill where El Faro, the highest lighthouse in the Americas, is located. Over 300 paved stone steps lead up to the base of El Faro where visitors can look out over the journey they've just taken. Be prepared with water and give yourself time for the climb.

The most amazing thing about a trip to Mexico is leaving knowing that you've only managed to see a small sliver of all that the country contains. Each trip offers a glimpse into lifetimes of history built upon gorgeous terrain. Mexico offers its beautiful landscapes full of high peaks, clear waters, and open sunsets. It offers its museums, its parks, its restaurants, and its performers. With so much to see packed into each state, and into each city, it is truly impossible to leave Mexico without finding the total eclipse vacation that you're seeking.

Hotels and Motels During the Eclipse

Once excitement of the total eclipse over Mexico spreads, rooms will become scarce. Many hotels in towns along the path of totality in the United States sold out for a year or more during the 2017 total eclipse. Mexico is not alone in this challenge. Hotels all along the path of totality will sell out in anticipation of the 2024 total eclipse.

What does this mean for eclipse visitors? Lodging and room rentals in eclipse towns will be at a massive premium. Does that mean all hope is lost to find a place to stay? Not at all. But you will have to be creative. There will be few, if any, hotel rooms available in these eclipse cities by early 2024. Accommodations in the cities and towns along the path of the eclipse will be difficult to come by.

In summer 2017, the author searched on Hotels.com for rooms along the 2017 total eclipse path on the weekend of August 21 and found many major cities sold out. Once word of the 2024 eclipse spreads, room rates will increase and availability will drop.

Search for rooms farther away from the eclipse path. If you are willing to stay in cities outside the eclipse path, you will have better success at finding rooms. As the eclipse approaches, people will book rooms farther from the totality path. By early spring, rooms in cities near the total eclipse path may be unavailable. The effect of this event will be felt across Mexico, the United States, and Canada.

Think regionally when looking for rooms. Be prepared to search far and wide during this major event. If a five-hour drive is manageable, your lodging options greatly expand, but it also increases your travel risk.

Internet Rentals

To find rooms to stay in towns along the eclipse path, try a web service such as Airbnb.com. Note that some people rent out rooms or homes illegally, against zoning regulations. Cities will feel the crunch of inquiries early due to others who experienced the 2017 eclipse.

If cities fully enforce zoning laws, authorities may prevent your weekend home rental. Online home rentals during the eclipse will be a target for rental scams. People from out of the area steal photos and descriptions, then post the home for rent. You send your check or wire money to a "rental agent" then show up to find you have been scammed. If the deal sounds strange or too good to be true, run away.

Camping

If you can book a campsite, do it as soon as you can. Do not wait. Many camping areas in Mexico are first-come, first-served. Major tourist points may be packed. Expect all areas to be swarming with people. Show up early to stake out your spot. Consider staying farther away and driving early on April 8. Camping in April around central Mexico will be comfortable if not slightly warm. You will need to be prepared for variable warm and cold desert conditions.

Please respect private land too. Property owners don't appreciate people overrunning their property without permission. In a massive country with millions of residents, people are very protective, but they're friendly, too. You never know what you might be able to arrange with a smile and a bit of money.

This all said, there are plenty of camping opportunities throughout Mexico. You don't have to sleep exactly on the eclipse path. If you're ready to rough it, there are camping options.

The management of visitors and traffic flow will be regulated by local and state governments. Be aware that rules may unexpectedly change without warning. Be prepared with contingency plans in case your initial plans are blocked.

National Parks and Monuments

Finding a camping site at any state park, local park, or protected reserve along the eclipse path in Mexico will be challenging. To watch the eclipse from any location, you do not have to sleep in it. You just need to drive there in the morning.

Law enforcement will be present on the eclipse weekend. Hundreds of thousands of people are expected in the region. Parking may overflow in popular cities. It will make parking lots and lines on major shopping days at the mall look uncrowded. For an event of this magnitude, find your location as early as possible.

The most important point to understand is you want to be at your viewing location well before the event. As the eclipse happens in the middle of the day across Mexico, you will have some time. However, you do not want to be caught unaware by weather, traffic, or other unexpected conditions.

Roadside camping (sleeping in your car) may be restricted by local authorities. Park facilities are only designed to handle so many people per day. Water, trash collection, and toilets can only withstand so much. If you notice trash on the ground, take a moment to throw it away. Be a respectful visitor and leave the area as clean as you found it.

Sleep in Your Car

Countless RVs, campers, trucks, cars, and travelers will flood toward the total eclipse path in Mexico. Sleeping in your car with friends is tolerable. Doing so with unadventurous spouses or children is another matter.

Do not be caught along the path of the total eclipse without some sort of plan, especially in the bigger cities of Mexico. The whole path of totality will fill with people on April 8.

Useful Local Webcams

Local webcams are handy to make last-minute travel decisions. The webcams are sensitive enough to show headlights at night. Use them to determine if there are issues before traveling out. Eclipse traffic will add to the morning commuter traffic.

The smartphone application Wunderground is useful to check on webcams in one place. Selected webcams are listed in the app. Whether you use this app or another, an Internet search will reveal many useful webcams for your search. Plan to find local webcams in Mexico well in advance of the event.

Weather

It's all about the weather during the eclipse. Nothing else will matter if the sky is cloudy. You can be nearly anywhere along the path in

Mexico and catch a view of the event when traffic comes to a standstill. But if there's a cloud cover forecast, seriously reconsider your viewing location.

Travel early wherever you plan to go. Attempting to change locations an hour before the eclipse due to weather will likely cause you to miss the event. Mexico country roads can be narrow and slow. The number of vehicles will cause unexpected backups.

Modern Forecasts

Use a smartphone application to check the up-to-date weather. Wunderground is a good application and has relatively reliable forecasts for the region. The hourly forecast for the same day has been rather accurate for the last two years. The below discussion refers to features found in the Wunderground app. However, any application with detailed weather views will improve your eclipse forecasting skills.

Cloud Cover Forecast

The most useful forecast view is the visible and infrared cloud-coverage map. Avoid downloading this app the night before and trying to learn how to read it. Practice reading them at home. It's imperative to understand how to interpret the maps early.

All cloud cover, night or day, will appear on an infrared map. Warm, low-altitude clouds are shown in white and gray. High-altitude cold clouds are displayed in shades of green, yellow, red, and purple. Anything other than a clear map spells eclipse-viewing problems.

To improve your weather guess, use the animated viewer of the cloud cover. It will give you a sense of cloud motion. You can discern whether clouds or rain are moving toward, away from, or circulating around your location.

Normal Mexico Weather Pattern

The expected weather in spring is possibly the best time for a total eclipse in Mexico. The dry season is well underway in April on the coast. Much of the path of the eclipse covers vast tracts of desert land, lending total eclipse viewing to be a relatively easy affair. Much of the weather flows from the central Pacific region and into the Mazatlán area. Past the western mountains, the skies will often be clear.

The common weather pattern in April is slightly chilly in the morning to warm in the afternoon. Passing cold fronts in spring can bring regular cloud cover and rain showers. One of the advantages of choosing Mexico over the United States and Canada for the total eclipse is the dry climate of the country's interior. The rainy season is during the summer so there is less risk of poor weather during the total eclipse.

Historically, coastal Mexico tends to have moderate cloud cover during April. Prepare to make adjustments. If anything other than clear skies are predicted, drive to other parts of Mexico or the United States

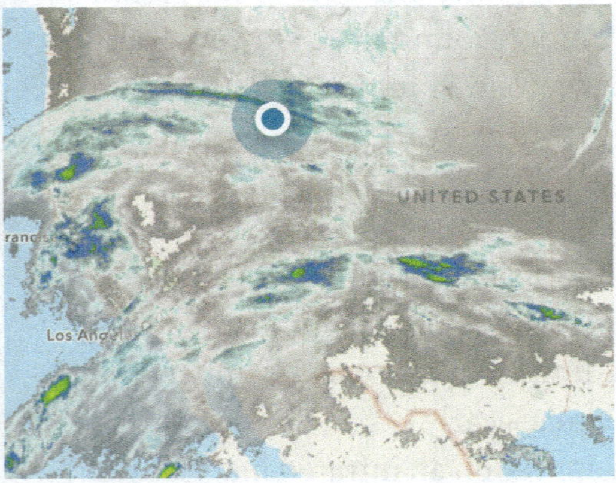

Infrared cloud map showing the worst case eclipse cloud cover.
Courtesy of National Weather Service.

Be aware of severe weather in Mexico. Although spring is well underway by the time of the total eclipse, there have been many substantial storms in April on the Mexican coast. Pay attention to the weather forecast. If dangerous weather is predicted, your main concern should be safety rather than chasing an eclipse.

Consider that slow-moving clouds can obscure the sun for far longer than the local duration of the totality. The time of totality is so short that you do not want to risk it. Missing it due to a single cloud will be a major disappointment for eclipse chasers.

Local Eclipse Weather Forecasts

Local town and city newspapers, radio, and television stations around Mexico will have a weekend edition with articles discussing the eclipse weather. However, conditions change unpredictably in Mexico. A three-day forecast for April may be incorrect.

Finding the Right Location to View Eclipse Effects

One of the peculiarities of total eclipses is that the entire show is not only in the sky. There are other unusual effects seen during the total eclipse that are worth looking for.

The first effect to watch for is the crescent moon shapes created from leaf shadows on the ground. They're best viewed on a sidewalk or asphalt. They can only be observed during the partial eclipse. The other effect that is worth watching for is the shadow bands or "snakes" as they're commonly called.

Shadow banding is seen right before and after the totality takes place. They're best observed on smooth, plain-colored surfaces. If you plan to be in the forest for the eclipse, you may struggle to see the bands but will likely see crescent shadows all around on the ground.

One of the challenges with all of these effects is choosing what to watch. You can see the crescent shadows in the hour before and after the totality but shadow banding happens before and after the totality. It is difficult to look away from the eclipse.

Road Closures and Traffic

Highways connecting various Mexican cities in the total eclipse path may be heavily impacted on the weekend before and day of the eclipse. As was found in the 2017 total eclipse, there is no way to predict which areas will be impacted.

Planning ahead is essential to give you the best opportunity to enjoy the eclipse without the nightmare of being stuck in traffic for hours on end. The traffic in the United States during the 2017 eclipse was stunning, so it may be substantial in Mexico in 2024.

Update yourself with the latest road report information from your local Mexican road condition website if it available in your location.

It's imperative to plan early and have one if not more backup plans in case of difficult travel conditions. April weather is unpredictable and variable.

If you believe it's necessary to leave a town to watch the eclipse, do so the night before or extremely early in the morning. RVs are common, and trains of them crawl through popular areas.

Communication Information

Cellular Phones

Cellular "cell" phone service in some remote locations may be problematic. Most of the time there is good coverage inside major cities and along roadways. However, even along major thoroughfares, there can be little or no coverage.

It's possible to find zones where text messages will send when phone calls are impossible. If you cannot make a phone call, the chance of having data coverage for web surfing or e-mail is low.

Please look up any information or communicate what you need before departing from the main roads around southeast Mexico. Some park areas sometimes have coverage. Planned to be self-contained. Plan for your cell phone not to connect.

You may find yourself out of cell service. With a large number of cell users in a concentrated area, coverage and data speed may collapse as well. Search on the phrase "cell phone coverage breathing".

Wilderness and Forest Safety

All Mexican mountains and wilderness areas are full of wild animals. Although beautiful, wild animals can be dangerous. They can easily injure or kill people, as they are far more powerful than humans. Do not try to feed any wild animals, including squirrels, foxes, and chipmunks, as they can carry diseases. These suggestions apply to all wild lands.

Ticks and Insects

Ticks and other insects exist all across Mexico, but not all species transmit disease. Ticks can transmit potentially life-threatening diseases such as Lyme disease. Other insects like kissing bugs transmit Chagas disease. Botflies, mosquitoes, and other insects can deliver many other unpleasant diseases to avoid if at all possible.

Spiders

Although the chance of encountering a venomous spider is low, it is not uncommon to encounter them. The two dangerous spiders in Mexico are the black widow and brown recluse. Should you encounter

either species, simply leave it alone. If you are bitten, seek immediate medical attention, as their toxin can be life-threatening.

Venomous Snakes

There are multiple species of venomous snakes in Mexico including the Querétaro dusky rattlesnake, banded rock rattlesnake, coral snake, Fer-de-lance, yellow-bellied sea snake, twin-spotted rattlesnake, and the Mojave green rattlesnake. Most of these snakes avoid humans. However, the Mojave green and the Fer-de-lance are highly aggressive. They can strike when provoked or threatened. Of the approximately 4,000 people annually bitten by venomous snakes in Mexico, approximately fifty people die according to the U.S. National Library of Medicine.

The best way to avoid encounters is to be mindful of your environment. Do not place your hands or feet in locations where you cannot clearly see the surroundings. Avoid heavy brush or tall weeds where snakes hide during the day. Step on a log or rock rather than over it, as a hidden snake might be on the other side. Rattlesnakes may not make any noise before striking.

Avoid handling all snakes. Should you be bitten, stay calm and seek medical attention as soon as possible. Transport the victim to the nearest mèdical facility immediately. Rapid professional treatment is the best way to manage snake bites. Refer to local authorities for more information on managing rattlesnakes injuries.

Travel Safety

Traveling to Mexico is rewarding and enjoyable. The total eclipse will be a unique experience for those in Mexico and for those who are visiting the country for the event. However, there are some travel dangers that all travelers must be aware of.

Food and water safety are some of the big risks that one has to face when traveling to any country. Unfamiliarity with the local customs of avoiding drinking tap water, consuming washed vegetables, and being mindful of how food is prepared can cause visitors distress. Daily tasks like using tap water to rinse a toothbrush and other seemingly innocuous tasks can bring real digestive distress to someone. Refer to your medical professional's advice when traveling to Mexico and how best to avoid these issues.

Being aware of one's surroundings and avoiding becoming an easy target of crime is critically important to one's safety. Although Mexico is portrayed in the media as a dangerous country, many of the areas of Mexico are statistically safer than the United States. Statistically, Mexico is as safe as the United States. However, some areas have more prevalent crime. Many of these problems are avoidable, though. Avoid traveling alone and after dark whenever possible. Refer to the latest U.S. Embassy & Consulates in Mexico travel advisories to find out what is going on at the time of the total eclipse at:

https://mx.usembassy.gov/mexico-travel-advisory/

https://travel.state.gov/

Renting vehicles in Mexico can be a great way to explore the country on your own. Be mindful of Mexico's laws regarding traffic and insurance. They are different than those in the United States.

Eclipse Day Safety

1. Hydrate

Spring temperatures are usually mild to warm. The excitement of the event can distract you from managing hydration. Drink plenty of water. Consume more than you would at home.

2. Eye Safety time

Use certified eclipse safety glasses at all times when viewing the partial eclipse. Only remove the glasses when the totality happens. Give your eyes time to rest. They can dry out and become irritated. Bring medically-approved eye drops to keep your eyes moist.

3. Sun exposure

Facing the sun for three hours can result in sunburns. Wear sunglasses and liberally apply sunscreen to avoid sunburns.

4. Eat well

Keep your energy up. Appetite loss is common when traveling. Maintain your normal eating schedule.

5. Prepare for temperature changes

Temperatures will drop rapidly during the eclipse and also once the sun sets. Bring appropriate clothing.

6. Talk with your doctor

If the humidity or heat bothers you talk with your doctor before traveling. Seek professional medical attention for serious symptoms.

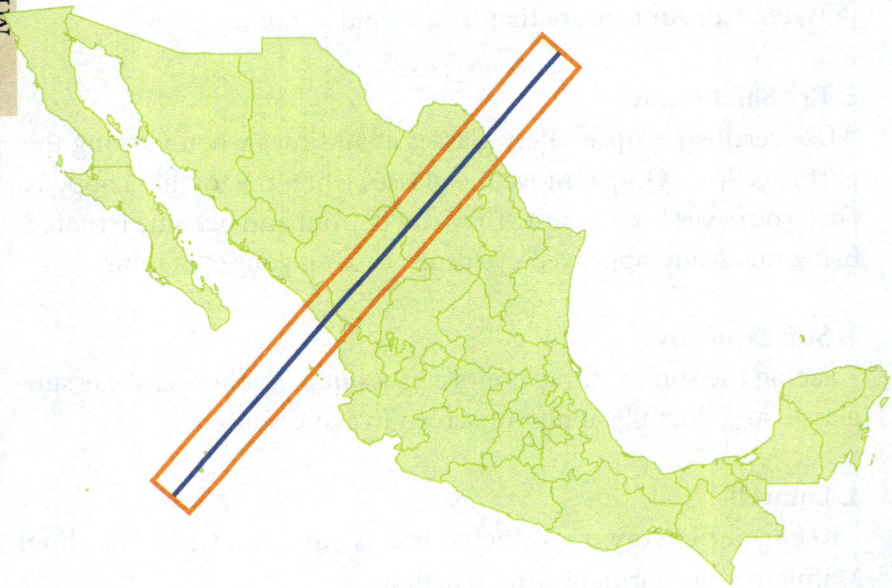

Total eclipse path across Mexico (approximate).

All About Eclipses

How an Eclipse Happens

An eclipse occurs when one celestial body falls in line with another, thus obscuring the sun from view. This occurs much more often than you'd think, considering how many bodies there are in the solar system. For instance, there are over 150 moons in the solar system. On Earth, we have two primary celestial bodies: the sun and the moon. The entire solar system is constantly in motion, with planets orbiting the sun and moons orbiting the planets. These celestial bodies often come into alignment. When these alignments cause the sun to be blocked, it is called an eclipse.

For an eclipse to occur, the sun, Earth, and moon must be in alignment. There are two types of eclipses: solar and lunar. A solar eclipse occurs when the moon obscures the sun. A lunar eclipse occurs when the moon passes through Earth's shadow. Solar eclipses are much more common, as we experience an average of 240 solar eclipses a century compared to an average of 150 lunar eclipses. Despite this, we are more likely to see a lunar eclipse than a solar eclipse. This is due to the visibility of each.

For a solar eclipse to be visible, you have to be in the moon's shadow. The problem with viewing a total eclipse is that the moon casts a small shadow over the world at any given time. You have to be in

* ILLUSTRATION NOT TO SCALE

a precise location to view a total eclipse. The issue that arises is that most of these locations are inaccessible to most people. Though many would like to see a total solar eclipse, most aren't about to set sail for the middle of the Pacific Ocean. In fact, a solar eclipse is visible in the same place on the world on average every 375 years. This means that if you miss a solar eclipse above your hometown, you're not going to see another one unless you travel or move.

It's much easier to catch a glimpse of a lunar eclipse, even though they occur at a much lower frequency than their solar counterparts. A lunar eclipse darkens the moon for a few hours. This is different than a new moon when it faces away from the sun. During these eclipses, the moon fades and becomes nearly invisible.

Another result of a lunar eclipse is a blood moon. Earth's atmosphere bends a small amount of sunlight onto the moon turning it orange-red. The blood moon is caused by the dawn or dusk light being refracted onto the moon during an eclipse.

Lunar eclipses are much easier to see. Even when the moon is in the shadow of Earth, it's still visible throughout the world because of how much smaller it is than Earth.

Total vs. Partial Eclipse

What is the difference between a partial and total eclipse? A total eclipse of either the sun or the moon will occur only when the sun, Earth, and the moon are aligned in a perfectly straight line. This ensures that either the sun or the moon is partially or completely obscured.

In contrast, a partial eclipse occurs when the alignment of the three celestial bodies is not in a perfectly straight line. These types of eclipses usually result in only a part of either the sun or the moon being obscured. This is often what led to ancient civilizations believing that some form of magical beast or deity was eating the sun or the moon. It appears as though something has taken a bite out of either the sun or the moon during a partial eclipse.

Total eclipses, rarer than partial eclipses, still occur quite often. It's more difficult for people to be in a position to experience such an event firsthand. Total solar eclipses can only be viewed from a small portion of the world that falls into the darkest part of the moon's shadow. Often this happens in the middle of the ocean.

The Moon's Shadow

The moon's shadow is divided into two parts: the umbra and the penumbra. The former is much smaller than the latter, as the umbra is the innermost and darkest part of the shadow. The umbra is thus the central point of the moon's shadow, meaning that it is extremely small in comparison to the entire shadow. For a total solar eclipse to be visible, you need to be directly beneath the umbra of the moon's shadow. This is because that is the only point at which the moon completely blocks the view of the sun.

In contrast, the penumbra is the region of the moon's shadow in which only a portion of the light cast by the sun is obscured. When

Total eclipse shadow 2016 as seen from 1 million miles on the Deep Space Climate Observatory satellite. Courtesy of NASA.

standing in the penumbra, you are viewing the eclipse at an angle. In the penumbra, the moon does not completely block the sun from view. This means that while the event is a total solar eclipse, you'll only see a partial eclipse. The umbra for the April 8 eclipse is over one hundred miles wide. The penumbra will cover much of the United States.

To provide some context, one total solar eclipse we experienced occurred on March 9, 2016, and was visible as a partial eclipse across most of the Pacific Ocean, parts of Asia, and Australia. However, the only place in the world to view this total solar eclipse was in a few parts of Indonesia.

Due to the varied locations and the brief periods for which they're visible, it's difficult to see each and every eclipse that occurs. Many people don't even realize that they have occurred. Consider that the umbra of the moon represents such a small fraction of the entire shadow and the majority of our planet is comprised of water. Thus, the rarity of being able to view a total solar eclipse increases significantly because it's likely that the umbra will fall over some part of the ocean rather than a populated landmass.

Eclipses Throughout History

Ancient peoples believed eclipses were from the wrath of angry gods, portents of doom and misfortune, or wars between celestial beings. Eclipses have played many roles in cultures, creating myths since the dawn of time. Both solar and lunar eclipses affected societies worldwide. Inspiring fear, curiosity, and the creation of legends, eclipses have cast a long shadow in the collective unconscious of humanity throughout history.

Early Myth & Astronomy

Documented observations of solar eclipses have been found as far back in history as ancient Egyptian and Chinese records. Timekeeping was important to ancient Chinese cultures. Astronomical observations were an integral factor in the Chinese calendar. The first observation of a solar eclipse is found in Chinese records from over 4,000 years ago. Evidence suggests that ancient Egyptian observations may predate those archaic writings.

Many ancient societies, including Roman, Greek and Chinese civilizations, were able to infer and foresee solar eclipses from astronomical data. The sudden and unpredictable nature of solar eclipses had a stressful and intimidating effect on many societies that lacked the scientific insight to accurately predict astronomical events. Relying on the sun for their agricultural livelihood, those societies interpreted solar eclipses as world-threatening disasters.

In ancient Vietnam, solar eclipses were explained as a giant frog eating the sun. The peasantry of ancient Greece believed that an eclipse was the sign of a furious godhead, presenting an omen of wrathful retribution in the form of natural disasters. Other cultures were less speculative in their investigations. The Chinese Song Dynasty scientist Shen Kuo proved the spherical nature of the Earth and heavenly bodies through scientific insight gained by the study of eclipses.

The Eclipse in Native American Mythology

Eclipses have played a significant role in the history of the United States. Before Europeans settled in the Americas, solar eclipses were important astronomical events to Native American cultures. In most native cultures, an eclipse was a particularly bad omen. Both the sun and the moon were regarded as sacred. Viewing an eclipse, or even being outside for the duration of the event, was considered highly taboo by the Navajo culture. During an eclipse, men and women would simply avert their eyes from the sky, acting as though it was not happening.

The Choctaw people had a unique story to explain solar eclipses. Considering the event as the mischievous actions of a black squirrel and its attempt to eat the sun, the Choctaw people would do their best to scare away the cosmic squirrel by making as much noise as possible until the end of the event, at which point cognitive bias

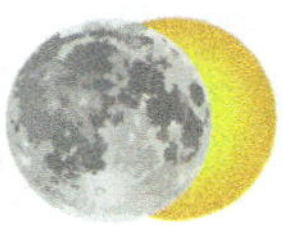

would cause them to believe they'd once again averted disaster on an interplanetary scale.

Contemporary American Solar Phenomena

The investigation of solar phenomena in twentieth-century American history had a similarly profound effect on the people of the United States. A total solar eclipse occurring on the sixteenth of June, 1806, engulfed the entire country. It started near modern-day Arizona. It passed across the Midwest, over Ohio, Pennsylvania, New York, Massachusetts, and Connecticut. The 1806 total eclipse was notable for being one of the first publicly advertised solar events. The public was informed beforehand of the astronomical curiosity through a pamphlet written by Andrew Newell entitled *Darkness at Noon, or the Great Solar Eclipse.*

This pamphlet described local circumstances and went into great detail explaining the true nature of the phenomenon, dispelling myth and superstition, and even giving questionable advice on the best methods of viewing the sun during the event. Replete with a short historical record of eclipses through the ages, the *Darkness at Noon* pamphlet is one of the first examples of an attempt to capitalize on the mysterious nature of solar eclipses.

Another notable American solar eclipse occurred on June 8, 1918. Passing over the United States from Washington to Florida, the eclipse was accurately predicted by the U.S. Naval Observatory and heavily documented in the newspapers of the day. Howard Russell Butler, painter and founder of the American Fine Arts Society, painted the eclipse from the U.S. Naval Observatory, immortalizing the event in *The Oregon Eclipse.*

Four more total solar eclipses occurred over the United States in the years 1923, 1925, 1932, and 1954, with another occurring in 1959. The October 2, 1959, solar eclipse began over Boston, Massachusetts. It was a sunrise event that was unviewable from the ground level. Eminent astronomer Jay Pasachoff attributed this event to sparking his interest in the study of astronomy. Studying under Professor Donald Menzel of Williams College, Pasachoff was able to view the event from an airline hired by his professor.

To this day, many myths surround the eclipse. In India, some local customs require fasting. In eastern Africa, eclipses are seen as a danger to pregnant women and young children. Despite the mystery and legend associated with unique and rare astronomical events, eclipses continue to be awe-inspiring. Even in the modern day, eclipses draw out reverential respect for the inexorable passing of celestial bodies. They are a reminder of the intimate relationship between the denizens of Earth and the universe at large.

Present Day Eclipses

The year 2017 brought the world's most-watched total eclipse in history on August 21, when a total solar eclipse crossed the United States. An annular eclipse, a "ring of fire," will pass over the United States in 2023 from Oregon to Texas. Though impressive, it will not compare to the 2024 total eclipse. There is little in nature that equals

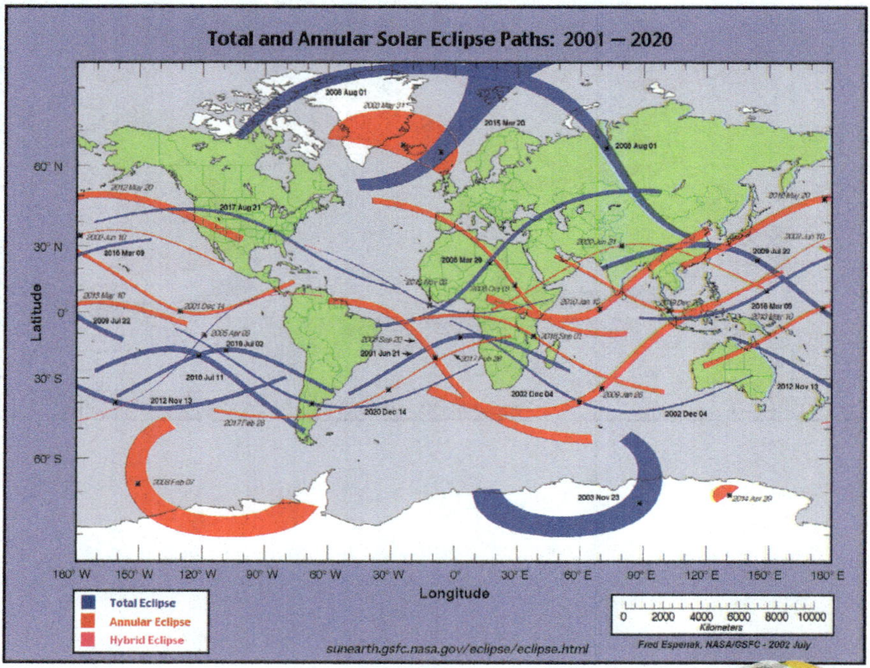

COURTESY OF NASA

the spectacle of the sun's corona and seeing stars in the day.

There will be multiple partial, annular, or hybrid eclipses across the world before the 2024 total eclipse. However many are in remote, inaccessible, or potentially dangerous locations on the globe. In 2019 and 2020, Chile and Argentina will experience total eclipses. The next total eclipse after that will occur over Antarctica in 2021. An extremely rare hybrid eclipse will happen in 2023 over the Indian Ocean, Australia, and Indonesia.

The next total solar eclipse viewable from the Mexico will occur on April 8, 2024. It will be visible in Mexico, the United States, and Canada.

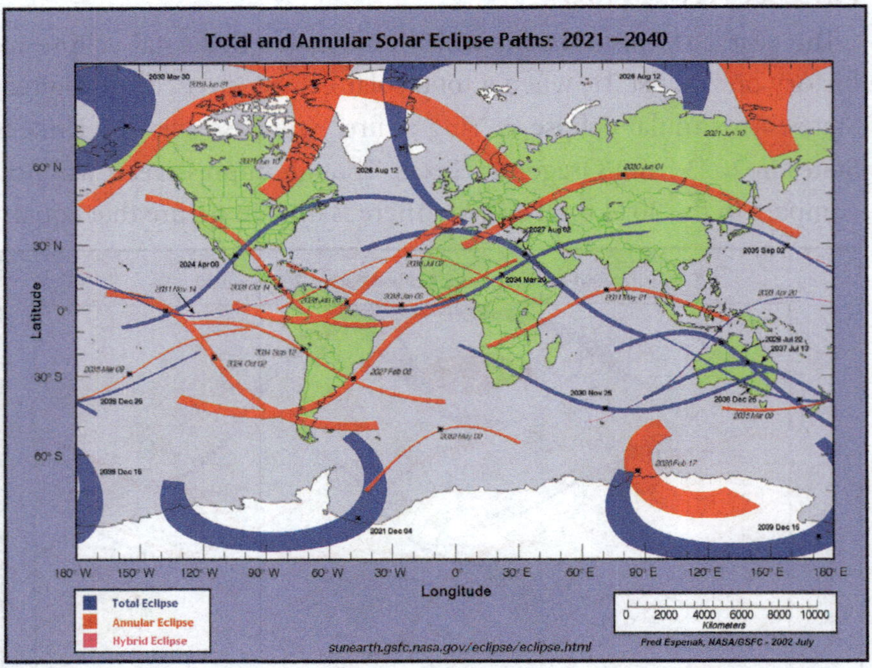

COURTESY OF NASA

Viewing and Photographing the Eclipse

At-home Pinhole Method

Use the pinhole method to view the eclipse safely. It costs little but is the safest technique there is. Take a stiff piece of single-layer cardboard and punch a clean pinhole. Let the sun shine through the pinhole onto another piece of cardboard. That's it!

Never look at the sun through the pinhole. Your back should be toward the sun to protect your eyes. To brighten the image, simply move the back piece of cardboard closer to the pinhole. To see it larger, move the back cardboard farther away. Do not make the pinhole larger. It will only distort the crescent sun.

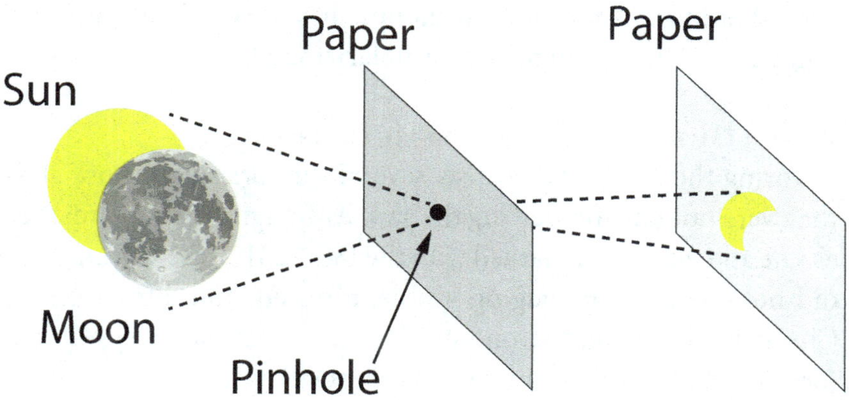

Welding Goggles

Welding goggles that have a rating of fourteen or higher are another useful eclipse viewing tool. The goggles can be used to view the solar eclipse directly. Do not use the goggles to look through binoculars or telescopes, as the goggles could potentially shatter due to intense direct heat. Avoid long periods of gazing with the goggles. Look away every so often. Give your eyes a break.

Solar Filters for Telescopes

The ONLY safe way to view solar eclipses using telescopes or binoculars is to use solar filters. The filters are coated with metal

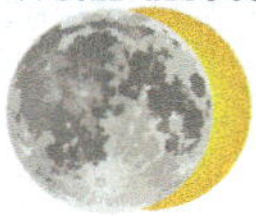

to diminish the full intensity of the sun. Although the filters can be expensive, it is better to purchase a quality filter rather than an inexpensive one that could shatter or melt from the heat.

The filters attach to the front of the telescope for easy viewing. Remember to give your telescope cooling breaks. Rapid heating can damage your equipment with or without filters attached.

Watch Out for Unsafe Filters

There are several myths surrounding solar filters for eclipse viewing. In order for filters to be safe, they must be specially designed for looking at a solar eclipse. The following are all unsafe for eclipse viewing and can lead to retinal damage: developed colored or chromogenic film, black-and-white negatives such as X-rays, CDs with aluminum, smoked glass, floppy disk covers, black-and-white film with no silver, sunglasses, or polarizing films.

Watch Out for Unsafe Eclipse Glasses

During the 2017 total eclipse, several vendors sold eclipse glasses that were not safe for viewing the sun. Although they were marketed as safe and were even marked with the ISO 12312-2 certification, they did not block eye-damaging visible, infrared, and ultraviolet light. Check the American Astronomical Society's website (eclipse.aas.org) for a list of reputable eclipse glasses vendors.

Viewing with Binoculars

When viewing the eclipse with binoculars, it is important to use solar filters on both lenses until totality. Only then is it safe to remove the filter. As the sun becomes visible after totality, replace the filters for safe viewing. Protect your pupils. Remember to give your binoculars a cool-down break between viewings. They can overheat rapidly from being pointed directly at the sun even with filters attached.

Planning Ahead

There are many things to keep in mind when viewing a total eclipse. It is important to plan ahead to get the most out of this extraordinary experience.

Understanding Sun Position

All compass bearings in this book are true north. All compasses point to Earth's magnetic north. The difference between these two measurements is called magnetic declination. The magnetic declination for central Mexico is:

6.68°E ± 0.34° (for Mazatlán in 2024)

Adjust the declination from the azimuth bearing as given in the text, and set your compass to that direction.

If you purchase a compass with a built-in declination adjustment,

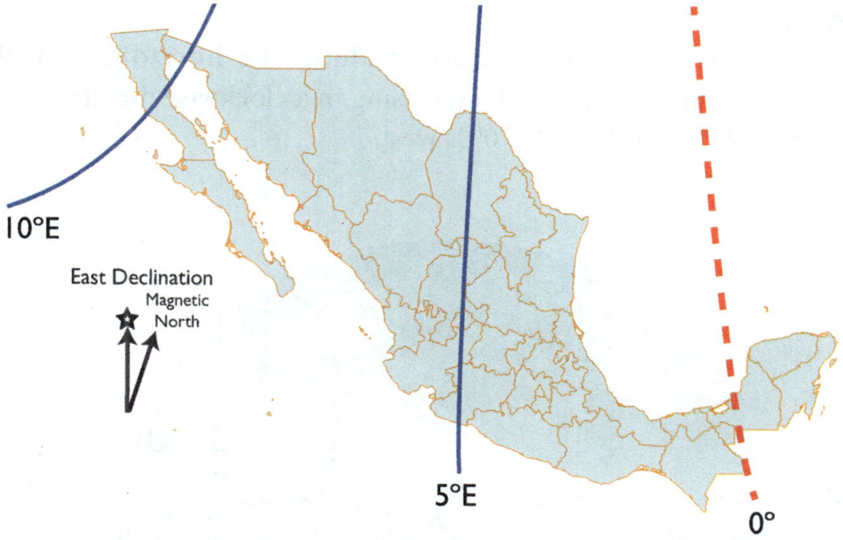

you can change the setting once and eliminate the calculations. The Suunto M-3G compass has this correction. A compass with a sighting mirror or wire will help you make a more accurate azimuth sighting.

The Suunto M-3G also has an inclinometer. This allows you to measure the elevation of any object above the horizon. Use this to figure out how high the sun will be above your position.

You can also use a smartphone inclinometer and compass for this purpose. Make sure to calibrate your smartphone's compass before every use, otherwise it might indicate the wrong bearing. Set the smartphone compass for true north to match the book. Understand the compass prior to April 8. There will be little time to guess or search on the Internet. Smartphone and GPS compasses are "sticky." Their compasses don't swing as freely as a magnetic compass does.

The author has used his magnetic compass for azimuth measurements and a smartphone to measure elevation. Combining these two tools will allow you to make the best sightings possible.

Outdoor sporting goods stores in most towns and cities carry compasses. Purchase and practice with a good compass in your hometown well before the event. Take the time to learn how to use it before the day of the eclipse. You do not want to struggle with orienteering basics under pressure.

Sun Azimuth

Azimuth is the compass angle along the horizon, with 0° corresponding to north, and increasing in a clockwise direction. 90° is east, 180° is south, and 270° is west.

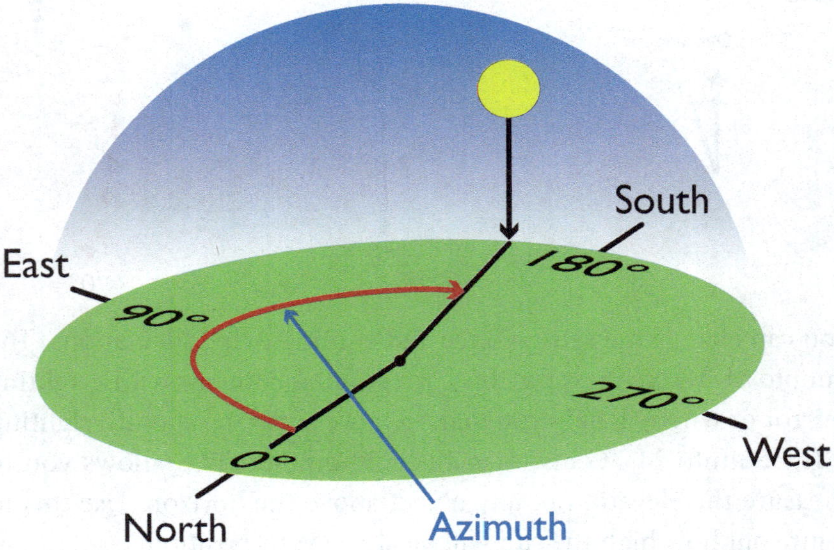

Sun Elevation

Altitude is the sun's angle up from the horizon. A 0° altitude means exactly on the horizon and 90° means "straight up."

Using the sun azimuth and elevation data, you can predict the position of the sun at any given time. Positions given in this book coincide with the time of eclipse totality unless otherwise noted.

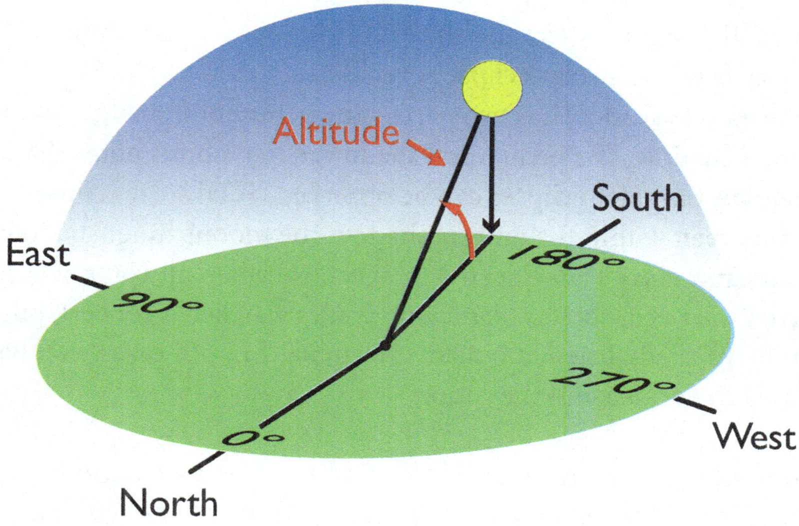

Eclipse Data for Select Mexico Locations

LOCATION	TOTALITY START (LOCAL TIME)
MELCHOR MÚZQUIZ	1:24:16PM
MONCLOVA	1:23:42PM
PIEDRAS NEGRAS	1:27:23PM
TORREÓN	1:16:55PM
DURANGO	1:12:07PM
MAZATLÁN	12:07:28PM

Eclipse Photography

Photographing an eclipse is an exciting challenge, as the moon's shadow moves near 2,570 KPH (1,600MPH). There is an element of danger and the pressure of time. Looking at the unfiltered sun through a camera can permanently damage your vision and your camera. If you are unsure, just enjoy the eclipse with specially designed eclipse glasses. Keep a solar filter on your lens during the eclipse and remove for the duration of totality!

Partial Vs. Total Solar Eclipse

To successfully and safely photograph a partial and total eclipse, it is important to understand the difference between the two. A solar eclipse occurs when the moon is positioned between the sun and Earth. The region where the shadow of the moon falls upon Earth's surface is where a solar eclipse is visible.

The moon's shadow has two parts—the penumbral shadow and the umbral shadow. The penumbral shadow is the moon's outer shadow where partial solar eclipses can be observed. Total solar eclipses can only be seen within the umbral shadow, the moon's inner shadow.

You cannot say you've seen a total eclipse when all you saw was a partial solar eclipse. It is like saying you've watched a concert, but in reality, you only listened outside the arena. In both cases, you have missed the drama and the action.

Photographing A Partial And Total Solar Eclipse

Aside from the region where the outer shadow of the moon is cast, a partial solar eclipse is also visible before a total solar eclipse within

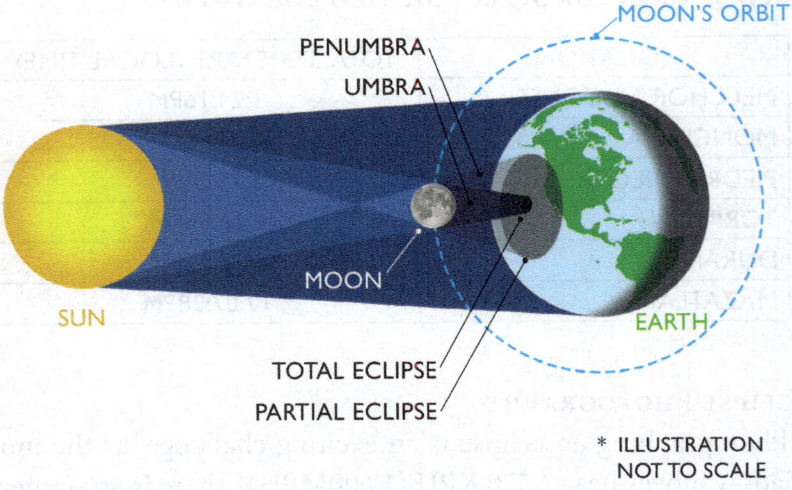

the inner shadow region. In both cases, it is imperative to use a solar filter on the lens for both photography and safety reasons. This is the only difference between taking a partial eclipse and a total eclipse photograph of the sun.

To photograph a total solar eclipse, you must be within the Path of Totality, the surface of the Earth within the moon's umbral shadow.

The Challenge

A total solar eclipse only lasts for a couple of minutes. It is brief, but the scenario it brings is unforgettable. Seeing the radiant sun slowly being covered by darkness gives the spectator a high level of anticipation and indescribable excitement. Once the moon completely covers the sun's radiance, the corona is finally visible. In the darkness, the sun's corona shines, capturing the crowd's full attention. Watching this phenomenon is a breathtaking experience.

Amidst all the noise, cheering, and excitement, you have no more than a few minutes to take a perfect photograph. The key to this is planning. You need to plan, practice, and perfect what you will do when the big moment arrives because there is no replay. The pressure is enormous. You only have a short time to capture the totality and the sun's corona using different exposures.

Plan, Practice, Perfect

It is important to practice photographing before the actual phenomenon arrives. Test your chosen imaging setup for flaws. Rehearse over and over until your body remembers what you will do from the moment you arrive at your chosen spot to the moment you pack up and leave the area.

You will discover potential problems regarding vibrations and focus that you can address immediately. This minimizes the variables that might affect your photographs at the most critical moment.

It's common for experienced eclipse chasers to lose track of what they plan to do. Write down what you expect to do. Practice it time and again. Play annoying, distracting music while you practice. Try photographing in the worst weather possible. Do anything you can to practice under pressure. Eclipse day is not the time to practice.

Once the sun is completely covered, don't just take photographs. Capture the experience and the image of the total solar eclipse in your mind as well. Set up cameras around you to record not just the total solar eclipse but also the excitement and reaction of the crowd.

Eclipse Photography Gear

What do you need to photograph the total eclipse? There are only a few pieces of equipment that you'll need. Preparing to photograph an eclipse successfully takes time. Not only do you have to be skilled and have the right gear, you have to be in the correct place.

Basic Eclipse Photography Equipment

- Solar viewing glasses (verify authenticity)
- Lens solar filter
- Minimum 300mm lens
- Stable tripod that can be tilted to 60° vertical
- High-resolution DSLR or mirrorless camera
- Spare batteries for everything
- Secondary camera to photograph people, the horizon, etc.
- Remote cable or wireless release

Additional Items

- Video camera
- Video camera tripod
- Quality pair of binoculars
- Solar filters for each binocular lens
- Photo editing software

Equipment to Prepare Before the Big Day

A. Solar viewing glasses

You need a pair of solar viewing glasses as the eclipse approaches.

B. Solar Filter

Partial and total eclipse photography is different from normal photography. Even if only 1% of the sun's surface is visible, it is still approximately 10,000 times brighter than the moon. Before totality, use a solar filter on your lens. Do not look at the sun with your eyes. It can cause irreparable damage to your retinas.

DO NOT leave your camera pointed at the sun without a solar filter attached. The sun will melt the inside of your camera. Think of a magnifying glass used to torch ants and multiply that by one hundred.

C. Lens

To capture the corona's majesty, you need to use a telescope or a telephoto lens. The best focal length, which will give you a large image of the sun's disk, is 400mm and above. You don't want to waste all your efforts by bringing home a small dot where the black disk and majestic corona are supposed to be.

D. Tripod

Bring a stable enough tripod to support your camera properly to avoid unsteady shots and repeated adjustments. Either will ruin your photos. It also needs to be portable in case you need to change locations for a better shot. ***Shut off camera stabilization on a tripod!***

E. Camera

You need to remember to set your camera to its highest resolution to capture all the details. Set your camera to:

- 14-bit RAW is ideal, otherwise
- JPG, Fine compression, Maximum resolution

Bracket your exposures. Shoot at various shutter speeds to capture different brightnesses in the corona. Note that stopping your lens all the way down may not result in the sharpest images.

Choose the lowest possible ISO for the best quality while maintaining a high shutter speed to prevent blurred shots. Set your camera to manual. Do not use AUTO ISO. Your camera will be fooled. The night before, test the focus position of your lens using a bright star or the moon.

Constantly double-check your focus. Be paranoid about this. You can deal with a grainy picture. No amount of Photoshop will fix a blurry, out-of-focus picture.

F. Batteries

Remember to bring fresh batteries! Make sure that you have enough power to capture the most important moments. Swap in fresh batteries thirty minutes before totality.

G. Remote release

Use a wired or wireless remote release to fire the camera's shutter. This will reduce the amount of camera vibration.

H. Video Camera

Run a video camera of yourself. Capture all the things you say and do during the totality. You'll be amazed at your reaction.

I. Photo editing software

You will need quality photo editing software to process your eclipse images. Adobe Lightroom and Photoshop are excellent programs to extract the most out of your images. Become well versed in how to use them at least a month before the eclipse.

J. Smartphone applications

The following smartphone applications will aid in your photography planning: Wunderground, Skyview, Photographer's Ephemeris, Sunrise and Sunset Calculator, SunCalc, and Sun Surveyor among others.

Camera Phones

Smartphone cameras are useful for many things but not eclipse photography. An iPhone 6 camera has a 63° horizontal field of view and is 3264 pixels across. If you attempt to photograph the eclipse, the sun will be a measly 30-40 pixels wide depending on the phone. Digital pinch zoom won't help here. If you want *National Geographic* images, you'll need a serious camera and lens, far beyond any smartphone.

Consider instead using a smartphone to run a time-lapse of the entire event. The sun will be minuscule when shot on a smartphone. Think of something else exciting and interesting do to with it. Purchase a Gorilla Pod, inexpensive tripod, or selfie stick and mount the smartphone somewhere unique.

Also, partial and total eclipse light is strange and ethereal. Consider using that light to take unique pictures of things and people. It's rare and you may have something no one else does.

Focal Length & the Size of Sun

The size of the sun in a photo depends on the lens focal length. A 300mm lens is the recommended minimum on a full-frame (FF) DSLR. Lenses up to this size are relatively inexpensive. For more magnification, use an APS-C (crop) size sensor. Cameras with these sensors provide an advantage by capturing a larger sun.

For the same focal length, an APS-C sensor will provide a greater apparent magnification of any object. As a consequence, a shorter, less expensive lens can be used to capture the same size sun.

The below figure shows the size of the sun on a camera sensor at various focal lengths. As can be seen with the 200mm lens, the sun is quite small. On a full-frame camera at 200mm, the sun will be 371 pixels wide on a Nikon D810, a 36-megapixel body. A lower resolution FF camera will result in an even smaller sun.

Printing a 24-inch image shot on a Nikon D810 with a 200mm lens at a standard 300 pixels per inch results in a small sun. On this size paper, the sun will be a miserly 1.25 inches wide!

Photographing the eclipse with a lens shorter than 300mm will leave you with little to work with. Using a 400mm lens and printing a 24-inch print will result in a 2.5-inch-wide sun. For as massive as the sun is, it is a challenge to take a large photograph of the sun. The sun will appear to move fast with a 500mm lens, too. Plan to adjust.

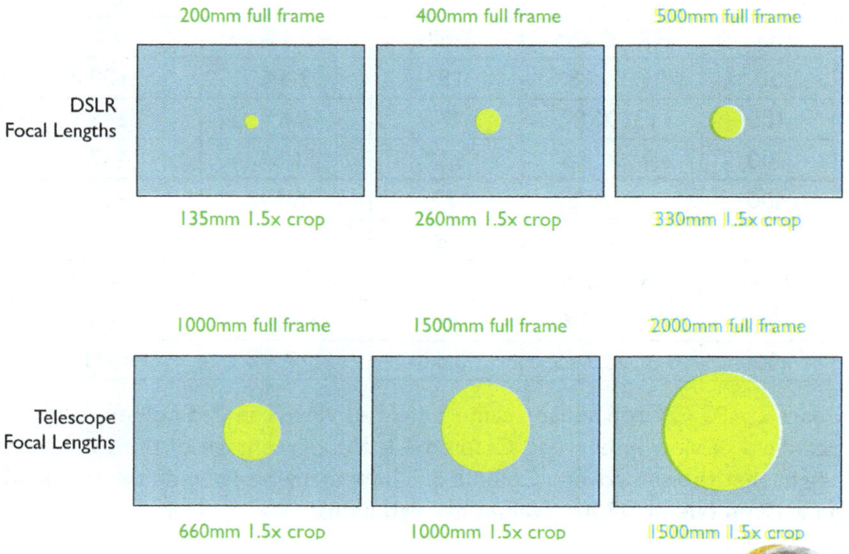

FOCAL LENGTH	FOV FULL FRAME	FF VERT. ANGLE	% OF FF	SUN PIXEL SIZE
14	104° X 81°	81°	0.7%	32.1
20	84° X 62°	62°	0.9%	41.9
28	65° X 46°	46°	1.2%	56.5
35	54° X 38°	38°	1.4%	68.5
50	40° X 27°	27°	2.0%	96.4
105	19° X 13°	13°	4.1%	200.2
200	10° X 7°	7°	7.6%	371.9
400	5° X 3.4°	3.4°	15.6%	765.6
500	4° X 2.7°	2.7°	19.6%	964.2
1000	2° X 1.3°	1.3°	40.8%	2002.5
1500	1.4° X 0.9°	0.9°	58.9%	2892.6
2000	1° X 0.68°	0.68°	77.9%	3828.4

Chart 1: Full-frame camera field of view. The 3rd column is the vertical field of view in degrees. Column 4 is the percentage of the total sensor height that the sun covers. Column 5 is how many pixels wide the sun will be on a 36MP Nikon D810. (Values are estimates)

PHOTOGRAPHY

FOCAL LENGTH	FOV CROP	CROP VERT DEG	% OF CROP	SUN PIXEL SIZE
14	80° X 58°	58°	0.9%	33.9
20	61° X 43°	43°	1.2%	45.8
28	45° X 31°	31°	1.7%	63.5
35	37° X 25°	25°	2.1%	78.7
50	26° X 18°	18°	2.9%	109.3
105	13° X 8°	8°	6.6%	245.9
200	6.7° X 4.5°	4.5°	11.8%	437.2
400	3.4° X 2°	2°	26.5%	983.7
500	2.7° X 1.8	1.8°	29.4%	1093.0
1000	1.3° X 0.9°	0.9°	58.9%	2186.0
1500	0.9° X 0.6°	0.6°	88.3%	3278.9
2000	0.6° X 0.45°	0.5°	117.8%	4371.9

Chart 2: APS-C Crop sensor camera field of view. The 3rd column is the vertical field of view in degrees. Column 4 is the percentage of the total sensor height that the sun covers. Column 5 is how many pixels wide the sun will be on a 12mp Nikon D300s. (Values are estimates)

The big challenge is the cost of the lens. Lenses longer than 300mm are expensive. They also require heavier tripods and specialized tripod heads. The 70-300mm lenses from Nikon, Canon, Tamron, and others are relatively affordable options. It is worth spending time at a local camera shop to try different lenses. Long focal-length lenses are a significant investment, especially for a single event.

To achieve a large eclipse image, you will need a long focal-length lens, ideally at least 400mm. A standard 70-300mm lens set to 300mm will show a small sun. At 500mm, the sun image becomes larger and covers more of the sensor area. The corona will take up a significant portion of the frame. By 1000mm, the corona will exceed the capture area on a full-frame sensor. See the picture in this chapter for sun size simulations for different focal lengths.

Suggested Exposures

To photograph the partial eclipse, the camera must have a solar filter attached. If not, the intense light from the sun may damage (fry) the inside of your camera. This has happened to the author. The exposure depends on the density (darkness) of the solar filter used.

As a starting point, set the camera to ISO 100, f/8, and with the solar filter on, try an exposure of 1/2000. Make adjustments based on the filter used, histogram, and highlight warning.

Turn on the highlight warning in your camera. This feature is commonly called "blinkies." This warning will help you detect if the image is overexposed or not.

Once the Baily's Beads, prominences, and corona become visible, there will only be a few minutes to take bracketed shots. It will take at least eleven shots to capture the various areas of the sun's corona and stars. The brightness varies considerably. No commercially available camera can capture the incredible dynamic range of the different portions of the delicate corona. This requires taking multiple photographs and digitally combining them afterward.

During totality, try these exposure times at ISO 100 and f/8: 1/4000, 1/2000, 1/1000, 1/250, 1/60, 1/30, 1/15, 1/4, 1/2, 1 sec, and 4 sec.

Disable camera/lens stabilization on a tripod!

Photography Time

Set the camera to full-stop adjustments. It will reduce the time spent fiddling. As an example, the author tried the above shot sequence, adjusting the shutter speed as fast as possible.

It took thirty-three seconds to shoot the above 11 shots using 1/3-stop increments. This was without adjusting composition, focus, or anything else but the shutter speed. When the camera was set to full stop increments, it only took twenty-two seconds to step through the same shutter speed sequence. Use a remote release to reduce camera shake.

Assuming the totality lasts less than two minutes, only four shot sequences could be made using 1/3-stop increments. Yet six-shot sequences could be made when the camera was set to full stop steps. Zero time was spent looking at the back LCD to analyze highlights and the histogram.

Now add in the bare minimum time to check the highlight warning. It took sixty-three seconds to shoot and check each image using full stops. And that was without changing the composition to allow for sun movement, bumping the tripod, etc. Looking at the LCD ("chimping") consumed **half** of the totality time.

This test was done in the comfort of home under no pressure. In real-world conditions, it may be possible to successfully shoot only one sequence. If you plan to capture the entire dynamic range of the totality, you must practice the sequence until you have it down cold. If you normally fumble with your camera, do not underestimate the difficulty, frustration, and stress of total eclipse photography.

Most importantly, trying to shoot this sequence allowed for zero time to simply look at the totality to enjoy the spectacle.

Avoid Last-Minute Purchases

You should purchase whatever you think you'll need to photograph the eclipse early. This event will be nothing short of massive. Remember the hot toy of the year? Multiply that frenzy by a thousand. Everyone will want to try to capture their own photo.

Do not wait until the last few weeks before the eclipse to purchase cameras, lenses, filters, tripods, viewing glasses, and associated material. Consider that the totality of the eclipse will streak across central Mexico. Everyone who wants to photograph the eclipse will order at the same time. If you wait until too late to buy what you need, it's conceivable that solar filters to create a total eclipse photo will be sold out in the United States. All filters sold out during the 2017 total eclipse. Whether this happens or not, do not wait to make your purchases. It may be too late.

Practice

You will need to practice with your equipment. Things may go wrong that you don't anticipate. If you've never photographed a partial or total eclipse, taking quality shots is more difficult than you think. Practice shooting the sequence with a midday sun. This will tell you if you have your exposures and timing correct. Figure out what you need well in advance.

Practice photographing the full moon and stars at night. Capture the moon in full daylight to learn how your camera reacts. Astrophotography is challenging and requires practice.

The August 21, 2017, eclipse as seen in Jackson, WY, shot with a Nikon D800 with an 80-400mm lens set to 340mm. The sun is 644 pixels wide on the 7360x4912 image.

This image is shown straight out of the camera without modification. Even with a high-quality camera and lens, photographing an eclipse is challenging.

The sun will follow this path similar to this on the day of the eclipse on April 8, 2024.

Note that this image is a simulation and approximation the sun's path and where the total eclipse may appear from one perspective. Refer to the eclipse position data for a more accurate location.

⊙ is the symbol for the sun and first appeared in Europe during the Renaissance.

☾ is the ancient symbol for the moon.

Viewing Locations Around Mexico

Tens of thousands of people will travel to and around Mexico to view the total eclipse. There are few obstructions and there is a vast amount of space to view the total eclipse from.

If the weather is questionable, seek out a new location as soon as possible. If you wait until the hour before the eclipse, you may find yourself stuck in traffic, as others will be looking for a viewing location. Be safe on the roadways, as drivers may be distracted.

This section contains popular, alternative, and little-known locations to watch the eclipse. As long as there are no clouds, the total eclipse will be viewable from any of these locations. For the nature locations, hiring a competent local guide is highly recommended.

Suggested Total Eclipse View Points

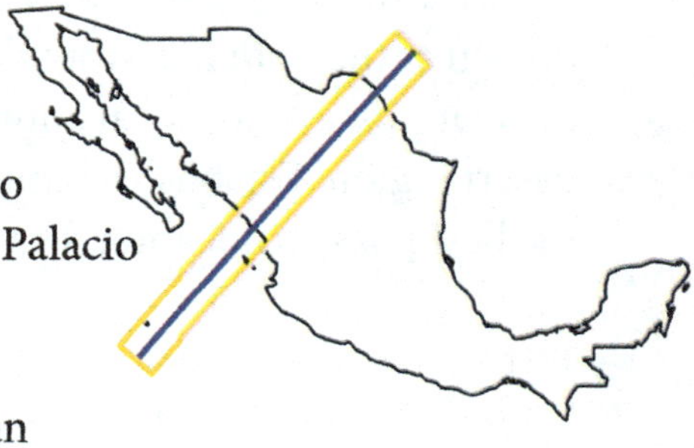

Towns and Cities

Coahuila
- Cuatro Ciénegas
- Melchor Múzquiz
- Monclova
- Piedras Negras
- Sabinas
- Torreón

Durango
- Durango
- Gomez Palacio
- Nazas

Sinaloa
- Mazatlán

Unique Locations

Coahuila
- Dunas de Yeso
- Sierra la Madera

Colima
- Isla Socorro

Durango
- Cañon de Fernández State Park
- Laguna de Santiaguillo
- El Tecúan Ecological Park
- Reserva de la Biósfera de Mapimí
- Zona Silencio

Sinaloa
- Lago Picachos
- Las Labradas
- Meseta de Cacaxtla

Cuatro Ciénegas

Elevation:	741m (2,433 feet)
Population:	12,150
State:	Coahuila

Overview

Cuatro Ciénegas promises to be one of the best jumping-off points for viewing the total eclipse in northern Mexico. This town is located near many popular destinations including the highly rare white dunes of Dunas de Yeso. Named after the marshy region, people have been settled in this area for thousands of years. The availability of water in an otherwise desert landscape makes it possible to support a community here. Interestingly, the streams and rivers flow into the area but have no known outlet, similar to the Great Basin region of the United States.

Take a drive or a tour to Poza Azul, the most popular and well-known of the wetlands around Cuatro Ciénegas. These water sources make it possible for local animals to thrive in an otherwise parched desert region. Establish a base camp here and then strike out to find the best possible location for viewing the total eclipse.

Totality Duration

4 minutes 14 seconds

Notes

https://www.visitmexico.com/en/coahuila-de-zaragoza/cuatro-cienegas

Event	Time (UTC-5)	Altitude	Azimuth
Sunrise	7:31:00am	0°	81°
Eclipse Start	12:04:35pm	58°	123°
Totality Start	1:21:48pm	69°	159°
Totality End	1:26:02pm	69°	162°
Eclipse End	2:46:04pm	66°	217°
Sunset	8:09:00pm	0°	279°

Melchor Múzquiz

Elevation:	485m (1,591 feet)
Population:	38,992
State:	Coahuila

Overview

Named after the fifth president of Mexico, Melchor Múzquiz, the city by the same name was named the seventh Pueblo Mágico (Magical Town) of Coahuila in 2018. This unique city is home to some of the most important paleontological finds in the region. A collection of dinosaur fossils is contained in the Museo de Paleontología (Museum of Paleontology). The fossils include those of sharks, crabs, and fish. The most valuable is the skeleton of a mosasaurus.

The city is also home to a thriving artisan community that takes advantage of the local veins of fluorite. These skilled artists create sculptures and furniture out of this phosphorescent mineral.

One of the important local cuisines is chorizo. This meat mixture is dried in the sun and then cured with spices and vinegar. It is popular and well worth trying.

Totality Duration

4 minutes 27 seconds

Notes

https://www.visitmexico.com/en/coahuila-de-zaragoza/melchor-muzquiz
https://www.turismocoahuila.com/melchor-muzquiz-turismo-coahuila/

Event	Time (UTC-5)	Altitude	Azimuth
Sunrise	7:28:00am	0°	81°
Eclipse Start	12:07:08pm	58°	125°
Totality Start	1:24:16pm	69°	163°
Totality End	1:28:43pm	69°	166°
Eclipse End	2:48:24pm	65°	218°
Sunset	8:07:00pm	0°	279°

Monclova

Elevation: 605m (1,984 feet)
Population: 215,271
State: Coahuila

Overview

Monclova is known as the steel capital of Mexico. This industrious city is the third-largest in Coahuila. The Museo Pape is well worth a visit, as it explains the city's industrial growth. The upper floors are dedicated to traveling exhibits that regularly feature Mexican history.

One of the industrial centers in Coahuila, Mexico. It is not far from the USA border and Texas. If you visit Monclova to experience the total eclipse, one of the best choices to enjoy the event is the beautiful Xochipilli I and II Parks. These surprising green parks in the otherwise desert landscape of Coahuila are a welcome relief. The 48 hectares (118 acres) split between the two parks has a beautiful fountain with a sculpture as well as waterfalls integrated into the landscape. Also, there is a zip line that is highly popular with children.

Totality Duration

2 minutes 20 seconds

Notes

http://www.museopape.org/
http://www.xochipilli.org.mx/

Event	Time (UTC-5)	Altitude	Azimuth
Sunrise	7:28:00am	0°	81°
Eclipse Start	12:05:21pm	59°	124°
Totality Start	1:23:42pm	69°	162°
Totality End	1:26:02pm	70°	164°
Eclipse End	2:47:04pm	66°	219°
Sunset	8:06:00pm	0°	279°

Piedras Negras

Elevation:	222m (728 feet)
Population:	150,178
State:	Coahuila

Overview

Piedras Negras, the city in Coahuila, Mexico, lies on the Rio Grande River on the United States and Mexican border. The city was founded in the middle of the nineteenth century. At the Grand Plaza, one of the largest Mexican flags in the country flies 120m (393 feet) over the city. This flag (at the time of this writing) is certified by the Guinness Book of World Records as the largest flying flag on a pole.

The Plaza de las Culturas has several large-scale models of the most important Mexican pyramids. These scale pyramids host many cultural events and are well worth visiting.

It is mostly known for the coal mines and two power plants. There are a lot of interesting sites in Piedras Negras. It is possible to walk across the Eagle Pass bridge from Texas to enter Mexico with the proper documentation. Piedras Negras is the last major city to see the total eclipse in Mexico.

Totality Duration

4 minutes 24 seconds

Notes

https://www.zocalo.com.mx/piedras-negras/

Event	Time (UTC-5)	Altitude	Azimuth
Sunrise	7:23:00am	0°	81°
Eclipse Start	12:10:06pm	59°	129°
Totality Start	1:27:23pm	68°	168°
Totality End	1:31:47pm	68°	171°
Eclipse End	2:51:15pm	63°	220°
Sunset	8:04:00pm	0°	279°

Sabinas

Elevation:	340m (1,115 feet)
Population:	47,993
State:	Coahuila

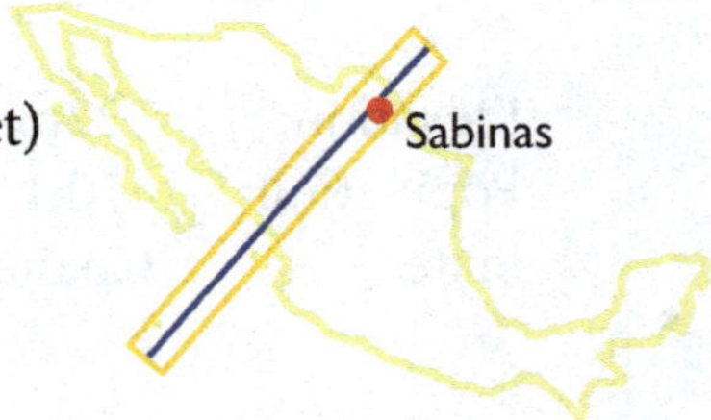

Overview

Sabinas, Coahuila, is one of the unique historical locations in northern Mexico. This small city contains a collection of unique artifacts at the Museo de Francisco Villa (Museum of Francisco Villa), also known as Lamar House. The museum is dedicated to famous revolutionaries Francisco Pancho Villa, and Sabinas is one of the critical historical places during the revolution. Pancho Villa and the Mexican government signed a peace treaty, here in Sabinas.

One very pleasant attraction in the city is Río Sabinas (Sabinas River). This large river winds its way along the southwest edge of the city. The sizable trees that line this river make for a pleasant getaway from the heat in the summer. An enterprising photographer will be able to create an original photograph with the trees, river, and total eclipse.

Totality Duration

4 minutes 12 seconds

Notes

http://sabinascoah.gob.mx/

https://www.turimexico.com/museo-francisco-villa-en-sabinas-coahuila

Event	Time (UTC-5)	Altitude	Azimuth
Sunrise	7:26:00am	0°	81°
Eclipse Start	12:07:38pm	59°	126°
Totality Start	1:25:00pm	69°	165°
Totality End	1:29:12pm	69°	168°
Eclipse End	2:49:02pm	64°	219°
Sunset	8:06:00pm	0°	278°

Torreón

Elevation:	1,128m (3,701 feet)
Population:	679,288
State:	Coahuila

Overview

The large city of Torreón is located in southern Coahuila on the border of the state of Durango. This second-largest city in Coahuila is the industrial and economic center of the state. The main site in Torreón is the statue Cristo de las Noas, one of the largest statues in Latin America. It is the second-largest depiction of Jesus Christ in the world (at the time of publication). The mix of the statue with the antenna towers will make for an otherworldly total eclipse photography experience.

One site that should not be missed is the Canal de la Perla, which used to be the way to bring water from the Nazas river to the Torreón's fields. This ever-expanding tourist attraction is for those who want to explore a subterranean world out of the desert heat. For those who like Hispanic culture, we recommend visiting the Museo Arocena (Arocena Museum). This modern museum is well worth the visit.

Totality Duration

4 minutes 9 seconds

Notes

https://www.visitmexico.com/en/coahuila-de-zaragoza/torreon

Event	Time (UTC-5)	Altitude	Azimuth
Sunrise	7:37:00am	0°	81°
Eclipse Start	11:59:55am	57°	118°
Totality Start	1:16:55pm	69°	151°
Totality End	1:21:04pm	70°	154°
Eclipse End	2:41:26pm	69°	213°
Sunset	8:13:00pm	0°	278°

Durango

Elevation: 1,880m (6,168 feet)
Population: 565,300
State: Durango

Overview

Victoria de Durango is located in Northern Mexico, and it was founded in the sixteenth century. For many years the city was called, simply, Durango. In the nineteenth century, the name is changed in Victoria de Durango, in honor of the first Mexican president Guadalupe Victoria, who was born here.

One of the unique locations around Durango is Tres Molinos. This rock escarpment and canyon has been used as a site for multiple films over the years. Located 31 km (19.2 miles) south of Durango on Highway 115, the possibilities for adventures in natural surroundings is only limited by imagination.

Totality Duration

3 minutes 47 seconds

Notes

http://visitdurango.mx/
http://visitdurango.mx/directorio-de-servicios-turisticos/canon-tres-molinos/

Event	Time (UTC-5)	Altitude	Azimuth
Sunrise	7:43:00am	0°	81°
Eclipse Start	11:55:14am	55°	113°
Totality Start	1:12:07pm	69°	143°
Totality End	1:15:54pm	70°	145°
Eclipse End	2:36:40pm	71°	209°
Sunset	8:17:00am	0°	278°

Gomez Palacio

Elevation: 1,139m (3,736 feet)
Population: 257,352
State: Durango

Overview

Gomez Palacio is known as a suburb of a larger city, named Torreón. It is located in the state of Durango. One of the most unique buildings in the city is Teatro Nazas. This upside-down pyramid building may be a unique opportunity to play with while photographing the total eclipse. The ridges and patterns created by the theater afford the enterprising photographer the chance to create an original image.

Gomez Palacio was founded in the nineteenth century, by a wealthy philanthropist Santiago Lavín, who donated part of his large hacienda to create the suburb in 1886. The land was a gift from Lavín for those who wish to be farmers.

The city was named after the governor Francisco Gomez Palacio. Today, Gomez Palacio is best-known for agriculture. Here, the farmers grew cotton, wine grapes, maze, and other vegetables and fruits.

Totality Duration

4 minutes 17 seconds

Notes

https://gomezpalacio.gob.mx/

Event	Time (UTC-5)	Altitude	Azimuth
Sunrise	7:37:00am	0°	81°
Eclipse Start	11:59:54am	57°	118°
Totality Start	1:16:48pm	69°	151°
Totality End	1:21:06pm	70°	154°
Eclipse End	2:41:22pm	69°	213°
Sunset	8:13:00pm	0°	278°

Nazas

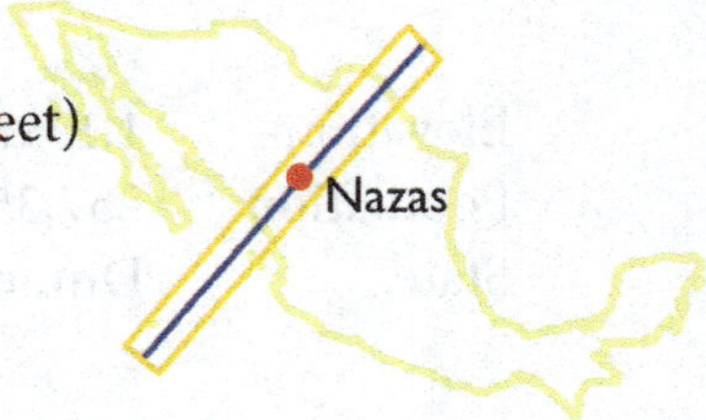

Elevation:	1,243m (4,078 feet)
Population:	3,622
State:	Durango

Overview

Nazas is a city in Mexican state Durango. It is a small town near the Nazas River. Nazas river is famous for an amateur kayak competition held in the summer. The Nazas river is also one of the longest rivers in Mexico. The high altitude combined with the cooler air of the river will make the air more stable, improving the chances of a perfectly clear sky for excellent eclipse photography.

The town of Nazas has the particular distinction of being at the apex of travel of the total eclipse. The sun will be at its highest point in the sky at Nazas during the event. This combine with the relatively high elevation of the town makes for the potential of the clearest possible photographs of the totality and corona.

Totality Duration

4 minutes 27 seconds

Notes

https://travelplanner.app/destinations/mexico/estado-de-durango/nazas/

Event	Time (UTC-5)	Altitude	Azimuth
Sunrise	7:40:00am	0°	81°
Eclipse Start	11:58:20am	56°	116°
Totality Start	1:14:59pm	69°	147°
Totality End	1:19:27pm	70°	150°
Eclipse End	2:39:39pm	70°	211°
Sunset	8:15:00am	0°	279°

Mazatlán

Elevation: 10 m (33 feet)
Population: 438,434
State: Sinaloa

Overview

Mazatlán lies on the west coast of Mexico in the state of Sinaloa. This city is the second-largest in the state and is a popular tourist destination for beach lovers. If you like specific architecture and wide beaches, Mazatlán is the place you must visit. People who wish to experience the total eclipse in the first large city in North America should travel to Mazatlán on April 8, 2024. Visit the Catedral Mazatlán Basílica de la Inmaculada Concepción in the historical Centro District. It will provide an interesting backdrop for a total eclipse photo.

One of the most popular viewpoints for the totality may be El Faro, a lighthouse situated on Cerro Creston, the highest hill in the city. Make sure to check with tour guides and government services well in advance if you plan this. As Mazatlán is an ideal city to start your Mexico total eclipse experience from, plan for hotels far in advance of the event.

Totality Duration

4 minutes 17 seconds

Notes

https://www.visitmexico.com/en/sinaloa/mazatlan

Event	Time (UTC-6)	Altitude	Azimuth
Sunrise	06:50:00am	0°	81°
Eclipse Start	10:51:25am	54°	110°
Totality Start	12:07:28pm	69°	135°
Totality End	12:11:45pm	69°	137°
Eclipse End	1:32:09pm	73°	201°
Sunset	7:24:00pm	0°	279°

Dunas de Yeso

Elevation: 764m (2,507 feet)
State: Coahuila

Overview

Located 27 km (16.7 miles), south of Cuatro Cienegas Dunas de Yeso (Gypsum Dunes) is one of only a few known gypsum dunes on Earth. These sugar-white dunes come from local gypsum deposits that have been pulverized over time and turned into soft sand. This 800-hectare (1,977 acres) otherworldly landscape may be one of the most interesting to view and photograph the total eclipse from.

One of the interesting parts of the dunes is the effect that the rare rainfall has on them. As gypsum is highly soluble in water, a crust can form over the surface after rain. This caked crust is fun to walk on in bare feet because it doesn't hurt and is more comfortable than regular sand to walk in. The experience of viewing the total eclipse from this rare location is something not to be missed.

Totality Duration

4 minutes 13 seconds

Notes

https://www.visitmexico.com/en/coahuila-de-zaragoza/cuatro-cienegas/dunas-de-yeso-gypsum-dunes
https://www.turismocoahuila.com/dunas-de-yeso

Event	Time (UTC-5)	Altitude	Azimuth
Sunrise	7:31:00am	0°	81°
Eclipse Start	12:04:10pm	58°	122°
Totality Start	1:21:23pm	70°	159°
Totality End	1:25:36pm	69°	161°
Eclipse End	2:45:40pm	66°	216°
Sunset	8:09:00pm	0°	279°

Sierra la Madera

Elevation: 3,040 m (9,973 feet)
State: Coahuila

Overview

Located 32 km (20 miles) west of Cuatro Cienegas, this highpoint of Coahuila is uniquely situated along the total eclipse path. Rising up above the valley floor at 1,890 meters (6,201 feet), this mountain prominence is only 3.3 km (2.1 miles) southeast of the centerline of the total eclipse.

One of the big advantages of viewing a total eclipse from the summit of a mountain is the ability to watch the moon's shadow sweep across the landscape. As the shadow moves at 2,570 KPH (1,600MPH) across the landscape, shadow banding may be visible depending on atmospheric conditions. This effect only lasts for moments and is well worth capturing on video and in still images. Be prepared and plan well because this moment is fleeting.

Totality Duration

4 minutes 27 seconds

Notes

https://www.peakbagger.com/peak.aspx?pid=13307

Event	Time (UTC-5)	Altitude	Azimuth
Sunrise	7:32:00am	0°	81°
Eclipse Start	12:04:15pm	58°	122°
Totality Start	1:21:15pm	69°	158°
Totality End	1:25:43pm	69°	161°
Eclipse End	2:45:36pm	66°	216°
Sunset	8:09:00am	0°	279°

Isla Socorro

Elevation: Sea level
State: Colima

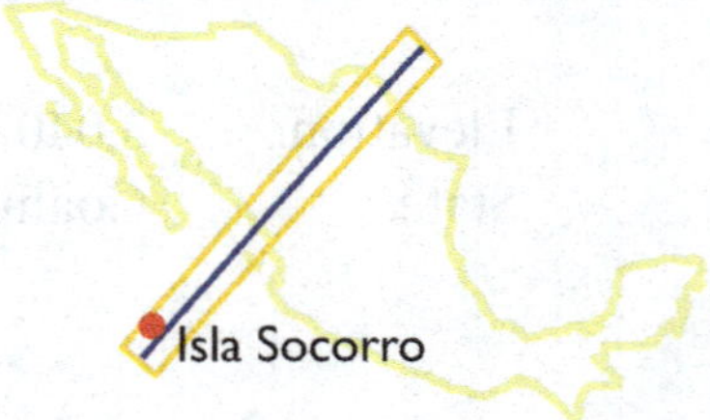

Overview

Island Socorro is a small volcanic island far off the Mexican coast. Located 600 kilometers (370 mi), this is the largest of the four islands of the Revillagigedo Archipelago. The volcano on this island is active and last erupted in 1993.

This island is a small paradise and is a bucket list destination for divers from around the world. The only way the general public can access this remote island is by a commercial charter. These are normally taken care of by diving companies. Most visitors stay on live-aboard dive boats during their time at the island. Contact and arrange a passage with a dive company as soon as possible. This will be a once-in-a-lifetime opportunity to enjoy the total eclipse from a natural paradise. The giant manta rays are famous for actively seeking out divers.

WARNING: Multiple sources list Isla Socorro in time zones UTC-5, UTC-7, and even UTC-8. Verify with the outfitter before departing. Adjust accordingly based on the UTC-5 chart listed below.

Totality Duration

3 minutes 19 seconds

Notes

https://www.padi.com/diving-in/socorro-island/

Event	Time (UTC-5)	Altitude	Azimuth
Sunrise	7:11:00am	0°	82°
Eclipse Start	11:37:14am	47°	99°
Totality Start	12:51:50pm	64°	113°
Totality End	12:55:09pm	65°	114°
Eclipse End	2:15:32pm	78°	167°
Sunset	7:39:00am	0°	278°

Cañon de Fernández State Park

Elevation:	1,174m (3,852 feet)
State:	Durango

Overview

Cañon de Fernández State Park is in the state of Durango, in northern Mexico. It represents the natural protected area in Chihuahuan dessert, and it covers 17,000 hectares (42,000 acres). Researchers have found bald cypress trees more than 1,400 years old in this special reserve area. These trees are named by the local indigenous people as *Ahuehuetl* meaning "old man in the water."

The park lies on the famous Nazas River. Beautiful landscapes, rocks, and trees provide visitors the opportunity to enjoy nature outside the bustling city of Torreón. The state park has been protected since 2004.

Totality Duration

4 minutes 16 seconds

Notes

https://www.visitmexico.com/en/durango/parque-canon-de-fernandez

Event	Time (UTC-5)	Altitude	Azimuth
Sunrise	7:38:00am	0°	81°
Eclipse Start	11:59:03am	56°	117°
Totality Start	1:15:55pm	69°	149°
Totality End	1:20:11pm	70°	152°
Eclipse End	2:40:31pm	69°	212°
Sunset	8:14:00pm	0°	278°

Laguna de Santiaguillo

Elevation: 1,958 m (6,424 ft)
Province: Durango

Overview

Located 111 km north of Durango, this is one of the only major bodies of water that will experience the total eclipse on April 8, 2024. This large body of water is an important refuge for the waterfowl that winter here. Birds migrate here from as far away as Alaska (US) and northern Canada. The lake is 24,000 hectares (59,300 acres), making it one of the largest in the state of Durango. The northern section of the lake has streams that feed it, keeping the lake permanent year-round. However, the southern section of the lake is seasonal. It only fills and floods when there are substantial rains in the rainy season.

This unique location will be a potentially good spot to enjoy the total eclipse from. The water from the lake will stabilize the surrounding air. This feature plus the combined high elevation will make it possible to take a series of clear, quality photographs of the total eclipse. Check with local guides and authorities for legal access to the lake shore.

Totality Duration

4 minutes 8 seconds

Notes

http://www.durangoturistico.com/destinos/santiaguillo.html

Event	Time (UTC-5)	Altitude	Azimuth
Sunrise	7:44:00am	0°	82°
Eclipse Start	11:56:45am	55°	114°
Totality Start	1:13:17pm	69°	144°
Totality End	1:17:26pm	69°	147°
Eclipse End	2:37:45pm	70°	208°
Sunset	8:18:00pm	0°	278°

El Tecúan Ecological Park

Elevation: 2,573m (8,446 feet)
State: Durango

Overview

Located 40 minutes southwest of the busy metropolis of Durango is El Tecúan, an ecological park in the middle of the high mountains of the region. The green landscape of this high desert stands in stark contrast to the regional landscape which is a dry desert. Cabins are available for rent and are worthwhile. As the reserve is in the mountains, the temperatures will drop significantly at night. Be prepared for chilly nights and warm days during the time around the total eclipse.

One of the major attractions to this particular park are the unexpected animals that can be seen here. Wolves, elk, and foxes are common sights in this special reserve.

Totality Duration

4 minutes 13 seconds

Notes

https://www.zonaturistica.com/en/things-to-do/durango/durango/ecological-park-the-tecuan
https://www.visitmexico.com/en/durango/el-salto/parque-ecologico-el-tecuan

Event	Time (UTC-5)	Altitude	Azimuth
Sunrise	7:45:00am	0°	81°
Eclipse Start	11:54:30am	55°	112°
Totality Start	1:11:02pm	69°	141°
Totality End	1:15:16pm	70°	143°
Eclipse End	2:35:47pm	71°	208°
Sunset	8:18:00am	0°	278°

Reserva de la Biósfera de Mapimí

Elevation: Varies
State: Durango

Overview

Situated at the intersection of the states of Durango, Coahuila, and Chihuahua, this large reserve is an excellent choice for watching the total eclipse. Be aware that the northern extent of the reserve is not in the path of totality, though. Make sure to travel to the southern section of the reserve to achieve the most time under the totality.

The Biósfera de Mapimí is on the list for UNESCO protection as a biosphere reserve. Mapimí Reserve is part of the Chihuahuan desert, and this is an excellent place to spend some time in nature. The reserve is the home of many animals and plants. Pumas, mule deer, and kit foxes are found in this region. This reserve is not simply a desert, though. There are many beautiful waterways and waterfalls to be found and enjoyed in this area.

Totality Duration

3 minutes 15 seconds

Notes

https://www.mexicodesconocido.com.mx/reserva-de-la-biosfera-de-mapimi-durango.html

Event	Time (UTC-5)	Altitude	Azimuth
Sunrise	7:38:00am	0°	81°
Eclipse Start	12:01:39pm	56°	119°
Totality Start	1:18:53pm	69°	153°
Totality End	1:22:07pm	69°	155°
Eclipse End	2:42:35pm	68°	212°
Sunset	8:15:00pm	0°	279°

LOCATIONS

Zona Silencio

Elevation: Varies
State: Durango

Overview

Nestled inside the Reserva de la Biósfera de Mapimí, the Zona Silencio (Silent Zone) is one of the strangest locations in the entire path of the April 8, 2024 total eclipse. This small region is famous for being the "Bermuda Triangle" of Mexico. It has been said that planes have lost communication radio signal, missiles have flown hundreds of kilometers off course to land here, and others have met unexpected strangers in the desert.

There is also an unexplained crater, cenote, or hundido in the area which is worth exploring. Locals have attributed it to all sorts of strange occurrences, from UFOs to glowing lights commonly seen in other desert regions of the world.

Totality Duration

2 minutes 57 seconds

Notes

https://mexicounexplained.com/zone-silence-mexicos-bermuda-triangle/
https://www.insolitaexperiencia.com/el-hundido-de-la-zona-del-silencio-portal-ovni/

Event	Time (UTC-5)	Altitude	Azimuth
Sunrise	7:37:00am	0°	81°
Eclipse Start	12:01:47pm	56°	119°
Totality Start	1:19:08pm	68°	153°
Totality End	1:22:05pm	69°	155°
Eclipse End	2:42:41pm	68°	211°
Sunset	8:14:00pm	0°	278°

Lago Picachos

Elevation: 92 m (303 feet)
State: Sinaloa

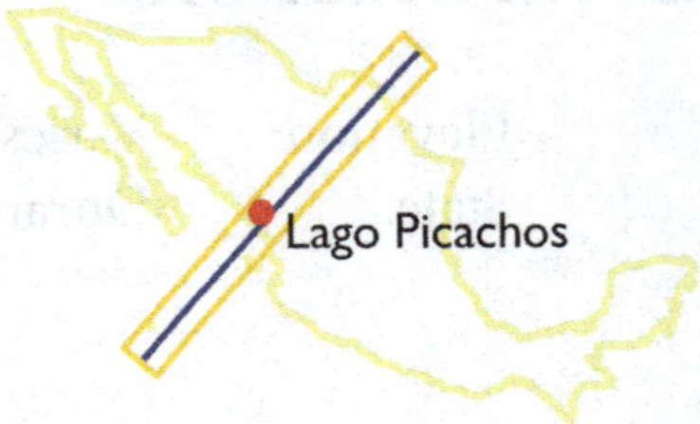

Overview

Located 48 km (30 miles) northeast of Mazatlán, Lago Picachos (Lake Picachos) is a reservoir that has become an attraction for anglers around the world. This allows visitors traveling to Mexico for the total eclipse to have a broader experience than just simply watching the eclipse. The cabins, resorts, and homes available to visitors around the lake are numerous with varying levels of amenities. Make sure to plan reservations far in advance of the total eclipse. As word of the event spreads, reservations will become scarce and those that remain will be expensive.

One of the advantages of viewing the total eclipse nearby or a lake is the stability of the air. Unlike the vast tracts of desert that the total eclipse passes over in Mexico, there are a few cooler options like Lago Picachos that will help improve eclipse photography.

Totality Duration

4 minutes 15 seconds

Notes

https://gomazatlan.com/do/lake-picachos/

Event	Time (UTC-6)	Altitude	Azimuth
Sunrise	6:49:00am	0°	81°
Eclipse Start	10:52:11am	54°	110°
Totality Start	12:08:19pm	68°	136°
Totality End	12:12:34pm	69°	138°
Eclipse End	1:32:58pm	72°	202°
Sunset	7:23:00pm	0°	278°

Las Labradas

Elevation: 17m (56 feet)
State: Sinaloa

Overview

Located 61 km (38 miles) northwest of Mazatlán on the coast, Las Labradas has made the tentative UNESCO Heritage Site list due to two features. The first is the amount of life that exists at this particular location. However, the true archaeological value is a large number of rock engravings from ancient peoples. The site has the highest concentration of rock carvings in Mexico and is the only known location of its type in a tidal zone.

Many of the rock engravings have been slowly eroded by the ocean action over the centuries. Interestingly, there is little known about the people who originally carved these rocks. The exact dates are a guess at best, from 1,000 B.C. to 300 A.D. The petroglyph work left by these unknown people of the region has left many mysteries that professional archaeologists still guess at today. It will be an interesting site to connect the total eclipse with a long-missing ancient culture.

Totality Duration

2 minutes 35 seconds

Notes

https://whc.unesco.org/en/tentativelists/5785/

Event	Time (UTC-6)	Altitude	Azimuth
Sunrise	6:51:00am	0°	81°
Eclipse Start	10:51:45am	53°	110°
Totality Start	12:08:28pm	68°	135°
Totality End	12:11:04pm	68°	136°
Eclipse End	1:32:09pm	73°	200°
Sunset	7:26:00pm	0°	278°

Meseta de Cacaxtla

Elevation: Varies
State: Sinaloa

Overview

Meseta de Cacaxtla is considered a premier destination for ecotourists in Sinaloa. Situated 60 km (37 miles) northwest of Mazatlán, this large protected habitat is home to reptiles, bird species, boars, and wild jaguars.

To try tasty food specialties, visit the La Chilacayota community. We can recommend Pan de Mujer. Cacaxtla Plateau is also an archaeological zone. Archaeologists found many petroglyphs (ancient engraved rocks) with different shapes from the pre-Columbian era. The Cacaxtla Plateau has over 50,000 hectares (123,000 acres) of protected land where communities watch over wildlife. Both tourists and researchers will be hard-pressed to find a better location for a natural surrounding for the total eclipse.

Totality Duration

2 minutes 51 seconds

Notes

https://www.mazatlantoday.net/mazatlan-meseta-cacaxtla-barras-piaxtla-day-trip-free-map.html

Event	Time (UTC-6)	Altitude	Azimuth
Sunrise	6:51:00am	0°	81°
Eclipse Start	10:52:01am	53°	110°
Totality Start	12:08:39pm	68°	135°
Totality End	12:11:31pm	68°	137°
Eclipse End	1:32:29pm	72°	200°
Sunset	7:25:00pm	0°	278°

Remember the Mexico Total Eclipse

April 8, 2024

Who was I with?

What did I see?

What did I feel?

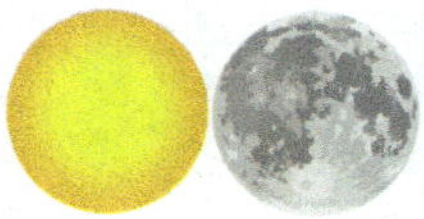

What did the people with me think?

Where did I stay?

Enjoy Other Books by Aaron Linsdau

50 Jackson Hole Photography Hotspots

This guide reveals the best Jackson Hole photography spots. Learn what locals and insiders know to find the most impressive and iconic photography locations in the United States. This is an excellent companion guide to the *Jackson Hole Hiking Guide*.

www.sastrugipress.com/books/50-jackson-hole-photography-hotspots/

Adventure Expedition One
by Aaron Linsdau M.S. & Terry Williams, M.D.

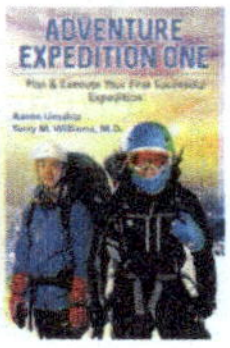

Create, finance, enjoy, and return safely from your first expedition. Learn the techniques explorers use to achieve their goals and have a good time doing it. Acquire the skills, find the equipment, learn to camp, understand medical issues, and learn the planning necessary to pull off an expedition.

www.sastrugipress.com/books/adventure-expedition-one/

Antarctic Tears

Experience the honest story of solo polar exploration. This inspirational true book will make readers both cheer and cry. Coughing up blood and fighting skin-freezing temperatures were only a few of the perils Aaron Linsdau faced. Travel with him on a world-record expedition to the South Pole.

www.sastrugipress.com/books/antarctic-tears/

How to Keep Your Feet Warm in the Cold

Keep your feet warm in cold conditions on chilly adventures with techniques described in this book. Packed with dozens and dozens of ideas, learn how to avoid having cold feet ever again in your outdoor pursuits.

www.sastrugipress.com/books/how-to-keep-your-feet-warm-in-the-cold/

Jackson Hole Hiking Guide

Jackson Hole contains some of the most dramatic and iconic landscapes in the United States. The book shares everything you need to know to hike Jackson's classic trails with canyons, high mountains, and hidden alpine lakes. This book is an excellent companion guide to *50 Jackson Hole Photography Hotspots.*

www.sastrugipress.com/books/jackson-hole-hiking-guide/

Subscribe to Aaron's YouTube channel at www.youtube.com/@alinsdau

If you enjoyed this book, please consider leaving a five-star review and a few words on what you liked about it at your favorite online retailer.

Lost at Windy Corner

Windy Corner on Denali has claimed fingers, toes, and even lives. What would make someone brave lethal weather, crevasses, and avalanches to attempt to summit North America's highest mountain? Aaron Linsdau shares the experience of climbing Denali alone and how you can apply the lessons to your life.
www.sastrugipress.com/books/lost-windy-corner/

The Motivated Amateur's Guide to Winter Camping

Winter camping is one of the most satisfying ways to experience the wilderness. It is also the most challenging style of overnighting in the outdoors. Learn 100+ tips from a professional polar explorer on how to winter camp safely and be comfortable in the cold.
www.sastrugipress.com/books/the-motivated-amateurs-guide-to-winter-camping/

Two Friends and a Polar Bear
by Terry Williams, M.D. & Aaron Linsdau

This story of friendship is about two old friends who plan to ski across the Greenland Ice Cap along the Arctic Circle in hopes of becoming one of the oldest teams to succeed.
www.sastrugipress.com/books/two-friends-and-a-polar-bear/

About the Author

Aaron Linsdau is the second American to ski alone from the coast of Antarctica to the South Pole (730 miles / 1174 km), setting a world record for surviving the longest expedition ever for that trip. He lead a 310-mile (499 km) ski expedition across the Greenland icecap along the Arctic Circle. Aaron has climbed Denali solo, crossed the Greenland tundra alone, skied across Yellowstone National Park solo, trekked through the Sahara desert, and successfully climbed Mt. Kilimanjaro and Mt. Elbrus in Russia.

Aaron Linsdau at the South Pole.

Use your smart device to scan the QR codes for website links.

Visit www.aaronlinsdau.com/subscribe to learn more about the author. Receive updates when he releases new books and shows.

Visit Sastrugi Press on the web at www.sastrugipress.com to purchase the above titles in bulk. They are available in print, e-book, or audiobook form.

Thank you for choosing Sastrugi Press.

Photography Notes

April 8, 2024

Photography Notes

April 8, 2024

Field and Travel Notes

April 8, 2024

Field and Travel Notes

April 8, 2024

Made in the USA
Monee, IL
03 November 2026

67381733R00046